Military Metallurgy

Military Metallurgy

ALISTAIR DOIG

Department of Materials and Medical Sciences
Cranfield University
The Royal Military College of Science
Shrivenham, UK

Routledge
Taylor & Francis Group

LONDON AND NEW YORK

FOR THE INSTITUTE OF MATERIALS

Book 696
First published in 1998
Reprinted with corrections in 2002

Maney Publishing
1 Carlton House Terrace
London SW1Y 5DB

ISBN 1–86125–061–4

Published 2015 by Routledge
2 Park Square, Milton Park, Abingdon, Oxon OX14 4RN
52 Vanderbilt Avenue, New York, NY 10017, USA

Routledge is an imprint of the Taylor & Francis Group, an informa business

CONTENTS

Preface and Acknowledgements

This book is an attempt to give a broad based view of metals in military service, covering several examples and rationales rather than just one or two in great depth. As such it is supposed to be informative and entertaining (sometimes maybe) rather than rigorously academic in its approach. For a start the title is strictly speaking incorrect since there are no 'air' or 'sea' examples, but "Army Metallurgy" does not have quite the same alliterative ring to it!

It is written for the militarist (who will hopefully appreciate the introductory metallurgy in the first three chapters) and for the metallurgist or materials scientist (who will I'm sure appreciate the introductory military technology encapsulated in all the chapters) and for the enthusiastic amateur alike. The content is based on some of the author's course notes compiled for undergraduate and post-graduate students at The Royal Military College of Science (RMCS), Shrivenham, most of whom are serving Army Officers.

After graduating in metallurgy at Leeds University the author worked at Stocksbridge steelworks, before going into contract research and then joining RMCS in 1975 to start lecturing. The semi-closed military area is not often met by most metallurgists (or even materiallurgists!) and there were many surprises in store - such as the use of 'temper embrittlement' in fragmenting steel shells, something that would be deliberately avoided in the civilian sector. Some of those surprises will now be shared with the reader.

I am most grateful to Harry Bhadeshia of Cambridge University for his encouragement to publish, and to Peter Danckwerts of The Institute of Materials for his editorial assistance. I am also indebted to Professors Alex Brown, Tony Belk and Cliff Friend for the facilities they have built up at RMCS, and to my many friends and colleagues in the Department of Materials and Medical Sciences who have helped me immensely over the years since joining RMCS Shrivenham. Last, but not least, I thank my mother and father for encouraging me to study metallurgy, and my wife Gem and sons James and Robert for their patience and support especially whilst writing this book.

Alistair Doig
April 1998

LIST OF PLATES

[all credits RMCS Shrivenham, except those stated in *itallics*]

1   Tensile test specimens and Charpy impact test specimen.
2   Tensile test machine. *Instron*
3   General purpose machine gun barrel GPMG - ductile fracture.
4   SS Schenectady – brittle fracture on a macro scale.
5   Charpy impact pendulum machine. *Avery*
6   Vickers hardness test machine. *Vickers*
7   Rockwell hardness test machine. *Avery*
8   Vickers hardness impression on cartridge brass.
9   Optical microscope *Reichart-Jung* ; Computerised image analyser.
10 Scanning electron microscope SEM. *JEOL*
11 Hardness gradient along the length of a 105 mm brass cartridge case.
12 105 mm brass disc, cup and finished case; Wrapped steel case.
13 60/40 brass microstructure.
14 70/30 brass microstructure – annealed at 650°C for 30 minutes.
15 70/30 brass microstructure – cold rolled 50% [CR].
16 70/30 brass microstructure –cold rolled 50% [CR] at higher magnification.
17 70/30 brass microstructure – CR then annealed at 350°C for 30 minutes.
18 70/30 brass microstructure – CR then annealed at 500°C for 30 minutes.
19 70/30 brass microstructure – CR then annealed at 750°C for 30 minutes.
20 Stress corrosion cracking SCC in 70/30 brass.
21 Mild steel cased ammunition round - 25 mm cannon.
22 Through-thickness section of shock loaded mild steel plate - scabbing.
23 76 mm and 105 mm steel projectile bodies –
    high explosive squash head HESH.
24 0.2%C steel microstructure – air cooled from 860°C.
25 0.4%C steel microstructure – air cooled from 860°C.
26 0.8%C steel microstructure – water quenched from 860°C.
27 0.8%C steel microstructure – water quenched from 860°C, then tempered
    at 550°C for 30 minutes.
28 SP 70 self-propelled 155 mm gun - with muzzle brake.
29 AS 90 self-propelled 155 mm gun. *VSEL*
30 SP 70 muzzle brake.
31 M107 SP 175 mm gun barrel.
32 Craze cracking on working surface of a 120 mm barrel section.
33 Craze cracking section – fatigue cracks growing from the rifling roots.
34 Microstructure of working surface of fired gun barrel - transverse section,
    optical micrograph.

35 Microstructure of working surface of fired gun barrel – transverse section, SEM micrograph.
36 Fracture of an old 'composite' wire wound 10" cannon barrel.
37 105 mm armour piercing discarding sabot kinetic energy penetrator round – APDS KE round – sectioned.
38 120 mm armour piercing fin stabilised discarding sabot kinetic energy penetrator round – APFSDS KE round.
39 120 mm APFSDS KE penetrator round – sabots separated.
40 Fired APFSDS soon after muzzle exit – sabots stripping away.
41 Microstructure of W-10%Ni,Fe penetrator alloy.
42 Microstructure of DU penetrator alloy.
43 Flash X-radiograph series – hydrodynamic penetration of a copper rod into an aluminium alloy target plate.
44 LAW 80 shaped charge anti-tank weapon system. *Hunting Engineering*
45 Mild steel target plates (each 25 mm thick) penetrated by a LAW 80 shaped charge jet. *Hunting Engineering*
46 Selection of copper shaped charge conical liners. *Hunting Engineering*
47 Flash X-radiograph of copper cone hydrodynamic collapse into a jet.
48 Experimental 120 mm tank launched shaped charge warhead.
49 Flash X-radiograph of copper jet penetrating hydrodynamically into an aluminium alloy target.
50 81 mm mortar.
51 81 mm mortar bomb body – cast iron.
52 Flake grey (automobile) cast iron microstructure.
53 Spheroidal graphite (sg) cast iron microstructure.
54 155 mm high explosive (HE) steel shell - fragmenting type.
55 Challenger main battle tank MBT – low alloy steel armour.
56 Through-thickness section of face hardened steel armour plate after small calibre KE attack.
57 Through-thickness section of steel plate penetrated by long rod KE – curvature of tract due to obliquity.
58 Armour failure by 'plugging' – macrosection (aluminium alloy).
59 'Gross cracking' of a 50 mm thick low alloy steel plate.
60 3%NiCrMo steel plate – through-thickness section microstructure.
61 3%NiCrMo steel plate – through-thickness section microstructure at higher magnification.
62 3%NiCrMo steel plate – section through fracture surface of through-thickness Charpy impact specimen, after testing at room temperature.
63 3%NiCrMo steel plate – SEM fractograph of through-thickness Charpy impact specimen, after testing at minus 196°C.
64 Electroslag remelted ESR 3%NiCrMo steel plate – through-thickness section microstructure.

1 Introduction to Metallurgy and Materials Selection

The science and technology of metals is diverse, covering aspects such as: extraction from ores, refining, alloying, castings and ingot production, primary production, secondary production to semi-finished products, heat treatment, quality control, mechanical property measurement, study of microstructures (using microscopes), atomic structure, materials selection, joining, machining, wear, corrosion, fatigue, environmental effects on mechanical properties, failure and fractography, and recycling of scrap. In this book, depending on the military example being discussed, some of these aspects will scarcely be mentioned - but the areas of **mechanical properties**, **microstructure** and **materials selection** keep recurring, and these are now introduced.

MECHANICAL PROPERTIES AND THEIR MEASUREMENT

Selecting the right materials is critical for the correct functioning of any engineering device, and this requires an understanding of their mechanical properties. The most common mechanical tests are now considered:

The Tensile Test to measure strength and ductility

A tensile specimen is dogbone shaped, either round or flat in section as seen in Plate 1. A flat specimen is shown here 'before' and 'after' testing. The central parallel portion, the 'gauge', is where most deformation occurs and lines are drawn to give the original **gauge length** L_o. The original **cross-sectional area** bearing the tensile force is A_o - the original specimen width times its thickness Wt in mm^2 units. After the test the broken two parts of the specimen are reconstituted to measure the final gauge length L and estimate ductility. A typical tensile test machine (tensometer) is shown in Plate 2.

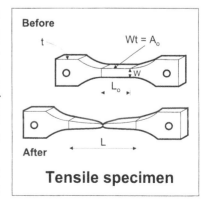

Before

t

$Wt = A_o$

W

L_o

After

L

Tensile specimen

The specimen heads are loaded into the tensometer grips and the specimen pulled to failure – usually at a crosshead speed of around 10 mm per minute (10 mm min⁻¹). Force is monitored by a **load cell** attached to one of the grips.

The X–Y recorder on the tensometer plots a force versus extension curve, which can then be rationalised to give a **tensile stress-strain curve**, so that values can be related to any size of component.

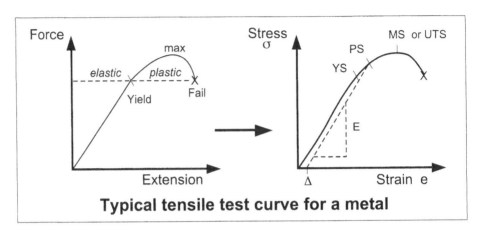

Typical tensile test curve for a metal

Engineering **stress** (σ) is *force/A$_o$* in N mm⁻², MN m⁻², or MPa units, and all three are numerically equivalent. Engineering **strain** (*e*) is *extension/L$_o$* which is dimensionless (mm/mm). These both relate to the original dimensions of the specimen which is very convenient. Sometimes true stress (*force/A*) and true strain 'ε' $[ln(L/L_o)]$ are used, A and L being instantaneous values requiring an extensometer to be attached to the specimen.

Initial loading is linear **elastic**, and this is reversible such that subsequent unloading will return the specimen to its original dimensions. The design engineer will usually try to choose a component cross-section such that the highest expected service stress is lower than the yield stress and by a reasonable safety factor. However, if the yield stress is exceeded then **plastic** or permanent deformation results. The engineering stress peaks at the maximum stress (MS) before dropping off to fracture, and this is due to localised 'necking' of the specimen - true stress climbs all the way to fracture.

Strength parameters measured in the tensile test are maximum stress **MS**, or ultimate tensile stress **UTS** as it is more usually called, and yield stress **YS**. Sometimes the limit of linearity is difficult to ascertain and a proof stress is measured instead by projecting an offset (Δ) up parallel to the elastic loading ramp - eg **0.2%PS**, where the offset is 0.2% of the gauge length. The offset can vary, usually between 0.1% and 2% of the gauge length, but all proof stress values are in excess of the yield stress.

Commercially pure aluminium would give tensile values of about 40 MPa YS and 90 MPa UTS, while ultra-high strength maraging steel would return values of around 2000 MPa YS and 2100 MPa UTS.

Tensile Stiffness or Young's modulus(E) is the slope of the elastic line, but this can be difficult to measure accurately because of test machine compliance. For metals,

Young's modulus varies from about 70 GPa for aluminium alloys to around 210 GPa for steels.

Ductility is defined as % elongation to fracture **%El** which is *100 x (L-L₀)/L₀*. A ductile alloy such as cartridge brass will give a value of about 65 %El. A thermosoftening polymer such as polythene can easily give a tensile ductility value of 500 %El. Most ceramics have very limited ductility (<2 %El) and their tensile properties have to be measured via bend testing.

Ductility can often be inferred from **fracture appearance.** For instance the burst **general purpose machine gun barrel** GPMG in Plate 3 reveals much plasticity, and so the heat treated low alloy steel used to make it is clearly fairly ductile. On the other hand the 'hogging' fracture of the hull of SS Schenectady in Plate 4 is macroscopically brittle - one can almost imagine weld repairing it in dry-dock without the need for much filler metal! This mode of failure was not uncommon in the **Liberty ships** of World War II as they crossed the Atlantic, often in winter making the likelihood of brittle fracture worse. They were amongst the first all-welded vessels, and grain growth in the weld heat affected zones HAZ was blamed. Afterwards the manganese content of weldable steels was increased to counteract this effect. The term **brittle** is used ambiguously by metallurgists. It is used to mean low ductility and also to mean low toughness, but the two are not always synonymous.

Toughness is defined as the energy to fracture E_f - units Nm or J. The area under the tensile curve is the energy to fail per unit gauge volume, and is a measure of toughness at slow **strain rate** - de/dt or $\dot{e}$, in mm/mm per second or s^{-1}. The initial strain rate is given by V/L_0 where V is the crosshead speed, and for a 20 mm gauge length pulled at a crosshead speed of 10 mm min⁻¹ this is about $8.10^{-3} s^{-1}$. However, in practice toughness is usually measured at higher strain rate, as in the impact test.

The Impact Test to measure comparative impact toughness

The most common impact specimen is the **Charpy** specimen, measuring 55 mm long by 10 mm square and with a 2 mm deep V-notch as a crack starter - seen in Plate 1 and drawn here. This is placed in the 40 mm gap in the anvil at the bottom of the Charpy pendulum machine shown in Plate 5, with the notch facing out. Then the raised

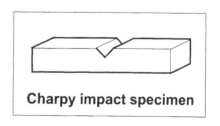

Charpy impact specimen

pendulum (with a tup mass of about 22 kg) is released to strike the specimen with 300 J energy at an impact speed of 5 ms⁻¹ - giving a strain rate at the notch root of around $3.10^2 s^{-1}$. The dial is calibrated to give a direct reading of energy to fracture (E_f) making the test quick, easy to perform, and ideal for quality control purposes. However, this test only gives **comparative impact toughness** values for specimens tested with this particular specimen geometry and in this particular way. For instance doubling the area of metal underneath the notch does not give twice the original E_f value, and altering the shape of the notch can cause the toughness 'league table' to change.

Charpy impact values for metals range from 1 J for grey cast iron to about 200 J for some quenched and tempered low alloy steels.

An **instrumented Charpy machine** has strain gauges fitted behind the striker tup making it possible to also measure the **force** acting on the specimen, and a force-time history is recorded on a transient recorder. This extra information is very useful towards a better understanding of the the whole fracture process. It can also be used to test fatigue pre-cracked specimens to measure **dynamic fracture toughness** (K_{Id}) from the peak force (P_Q). This parameter is geometry independent, giving an **absolute** measure of dynamic toughness. Fracture toughness is discussed further in the next section.

The Fracture Toughness Test - resistance to sharp crack propagation

There are two main types of specimen for this test - the single edge notch **SEN** specimen is similar to a large Charpy impact specimen but is tested in slow three-point bend mode (in the tensometer, reversed for compression), and the compact tension specimen **CTS** which is tested in tensile mode:

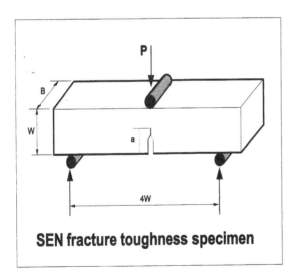

SEN fracture toughness specimen

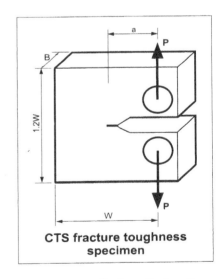

CTS fracture toughness specimen

Firstly, a **fatigue crack** is grown from the notch root by controlled cyclic loading, giving a consistent and sharp crack in every test . Then a clip gauge is fitted to the notch mouth to measure 'crack opening displacement' (to check there is no undue plasticity ahead of the crack) and the specimen is loaded at normal crosshead speed to fracture. After fracture the 'critical stress intensity factor' for final crack propagation K_Q can be calculated from the peak load P_Q and specimen geometry.

Lastly, a test validity checklist has to be satisfied and then the K_Q value becomes a valid K_{Ic} value (at last!).

K_{Ic} is the **fracture toughness** of the specimen in MPa m$^{1/2}$ units, and values for metals range from around 20 MPa m$^{1/2}$ for an as-cast magnesium alloy to about 200 MPa m$^{1/2}$ for a quenched and tempered low alloy steel.

Fracture toughness is an absolute material parameter (rather than being comparative like Charpy impact toughness) and can be directly used in stress analysis calculations on any size of component - provided the component is large enough.

The specimens in the above diagrams can be of different sizes, but their dimension ratios, as detailed in the test standard (British Standard BS 7448), must remain the same. It is important for the specimen breadth to be larger than a certain size depending on the material, and this is included in the validity checklist. It is not uncommon to find out at the end of the test that the specimen was too thin (K_Q does not then give a valid K_{Ic} value) and a second test is then required on a broader specimen.

The Hardness Test to measure resistance to indentation

A small area on a component or sample is polished with emery paper and an indenter applied under standard load and dwell-time conditions. This results in a surface impression, which is larger in a softer metal and smaller in a harder metal. The hardness impression is then sized under an optical microscope and this measurement converted into a hardness number. There are three hardness scales common in metallurgy, but fortunately all of them (and the geologist's Moh scale) are easily inter-related via tables:

The **Vickers** hardness machine, seen in Plate 6, uses an inverted pyramid shaped diamond indenter, as drawn right. This test gives H_V numbers (or VPN - Vickers pyramid numbers) and these are in kgf mm^{-2}, the load applied divided by the impression surface area, but the units are rarely quoted.

The **Brinell** hardness machine uses a hardened steel ball indenter, giving H_B numbers.

The **Rockwell** hardness machine (American in origin), seen in Plate 7, gives H_R numbers, in three

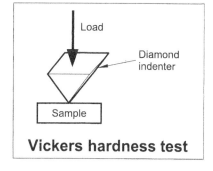

Vickers hardness test

scales A, B and C according to indenter type and load. Plate 8 is a micrograph of a Vickers hardness impression on a cartridge brass sample, taken at magnification X70. The sample was etched in acidified ferric chloride to also show the grain structure of the alloy - more on this later.

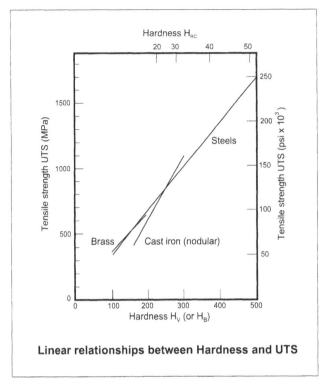

Linear relationships between Hardness and UTS

Commercially pure aluminium measures about 25 Hv and hardened steel can measure up to 800 Hv or so, with diamond itself estimated to be around 3500 Hv.

Hardness testing is simple to do, inexpensive and non-destructive.

An added bonus is that for metals there is a linear relationship between hardness number and tensile strength (UTS), making it extremely useful in quality control.

Other Tests

The tensile test, impact test, fracture toughness test and hardness test are the most commonly met, but other mechanical tests on materials include: fatigue (effect of cyclic loading on failure stress), creep (effect of high temperature), compression, shear, torsion, corrosion, and wear resistance.

These tell us the 'what' but the all important 'why' is obtained from the microstructure. By studying the microstructural features associated with say higher strength or higher toughness, these can hopefully be intentially designed into the next generation – and this approach is a major cornerstone of materials science and engineering.

MICROSTRUCTURE

It is often a surprise for newcomers to discover that metals are composed of **grains**, but on reflection most people can recall seeing the grain structure of galvanised steel products such as buckets or wheelbarrows. These grains originate as the nuclei of the solidifying metal, growing until they touch one another. The grain size of electroplated or hot-dipped zinc coatings is large enough to be seen with the naked eye (the

macrostructure) but bulk metals are usually mechanically worked and/or heat treated, which refines the as-cast grain structure and a microscope is needed to study the **microstructure**. A metallurgical **optical microscope** is seen in Plate 9, together with a CCTV computerised image analyser. A reflection microscope (rather than transmission) is needed to study opaque metals. A **metallography** specimen (or a 'micro' as it is often called) is prepared by sectioning, polishing to a mirror finish, and then etching chemically to reveal the microstructure.

This **micrograph** is of iron etched in 2% nitric acid in ethanol (2% nital) showing an **equiaxed** grain structure. If the iron had been plastically deformed then the grains would have been elongated in the direction of working. Some of the small dots seen are diamond paste particles embedded during the polishing, and others are **non-metallic inclusions** - impurities from the ingot stage of production. Average grain size here is about 80 microns (µm) - **a micron is a thousandth of a millimetre**. All other factors being equal, a finer grain size promotes both higher yield strength and higher ductility.

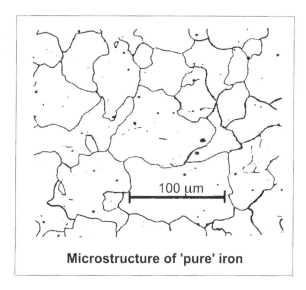

Microstructure of 'pure' iron

Using optical microscopy we are stuck with a planar 2-d section, so that some of the grains are sectioned through their polar caps while others are sectioned through their equators. There has to be a natural spread in 3-d grain size for them to fit together without voids, but this **sectioning effect** causes additional apparent variation. This is lived with for equiaxed microstructural features, but for directional structures it is common to examine more than one plane - for example longitudinal and transverse sections are often taken from bar samples.

An **electron microscope**, such as the scanning electron microscope **SEM** seen in Plate 10, is necessary for magnifications higher than about X1500. Increased resolution is obtained by using electrons rather than light and then magnifications in excess of X100,000 are possible. As well as higher magnification the SEM is capable of greater depth of focus, which is particularly useful when studying fracture surfaces (**fractography**). Also useful is the ability to carry out in-situ chemical analysis by energy dispersive analysis of X-rays **EDAX**. The electrons striking the specimen surface cause characteristic X-rays to be emitted, and their energy spectrum is analysed with a spectroscope attachment. If desired an area-scan can be used to plot out a map of local chemical analysis variations, such as microsegregation of alloying elements in a cast alloy. The electron beam can also be focussed onto small areas of the microstructure to allow in-situ **microanalysis**, which is particularly valuable when carrying out diagnostic fractography for instance.

In the transmission electron microscope **TEM** the electrons pass through a **thin foil specimen**, allowing direct observation of the smallest internal microstructural features. Then techniques such as **electron diffraction** can be used for analysis of the crystal lattice structure. It is perhaps a sobering thought that aluminium airframe alloys (among others) depend on sub-micron size precipitate particles within the grains for three-quarters of their yield strength, and these are impossible to study in the optical microscope.

MATERIALS COMPARISON AND SELECTION

Metals usually have a good compromise of strength, ductility, and toughness. A weak metal is often soft and ductile, whereas a high strength metal is harder and less ductile. These properties can be varied to suit the desired application by controlling the microstructure via alloy design, processing, and heat treatment.

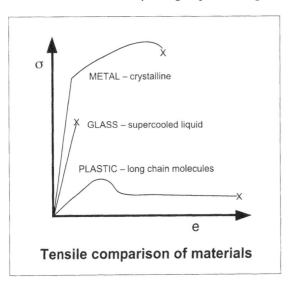

Tensile comparison of materials

Due to fundamental differences in atomic bonding, **non-metals** do not have the same combination of elasticity/plasticity:

Ceramics such as glass are very brittle, but have high melting points and good resistance to oxidation.

Thermosoftening plastics are weak with low stiffness, very ductile and easily formed to shape, but suffer from stress relaxation at room temperature and are not suitable for service at high temperatures.

Composite materials, such as glass fibre reinforced polymerics **GFRP** and carbon fibre reinforced polymerics **CFRP**, are an attempt to combine the best of these non-metal characteristics in single components.

Most non-metals are poor conductors of heat and electricity (which may or may not be advantageous), they do not suffer from corrosion, and they are usually less dense than metals. Some polymers swell and shrink according to humidity, and many suffer from degradation in the presence of organic solvents, and even embrittlement in the presence of UV light. Great strides are being made towards improving the toughness of ceramics and also towards improving the strength and stiffness of polymers and composites.

'**Work Hardening**' **occurs in metals** because of dislocations in their crystalline atomic structure. If a component is overloaded in service to above its yield stress YS, then during subsequent unloading **elastic recovery** occurs back down parallel to the elastic line. If there is a second overload the elastic limit is raised to YS[1]. This built-in active response to accidental overloading is often forgotten when designers change from metals to non-metals. Some polymers exhibit crystalline changes during necking, but this is not quite the same thing since 'drawing' then takes place in the unchanged material on either side of the neck.

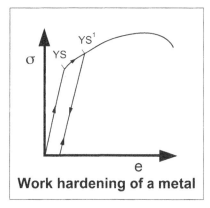

Work hardening of a metal

All of these factors and more (including price) need considering during the process of materials selection for particular engineering applications. It is very common to compare materials using **Tables of Mechanical Properties** such as that in the appendix (page 88) and other similar but more thorough compilations available elsewhere.

It is often enlightening to use **Ashby diagrams** where one property is graphed against another. The one here shows very clearly that most engineering ceramics, for example, exhibit superior '**specific stiffness**' (stiffness to weight ratio) compared to most metals - though they are so brittle that the tensile Young's modulus has to be calculated from the measured compression (bulk) modulus. A more detailed version of this diagram, and four other Ashby diagrams are in the appendix - pages 89 to 93.

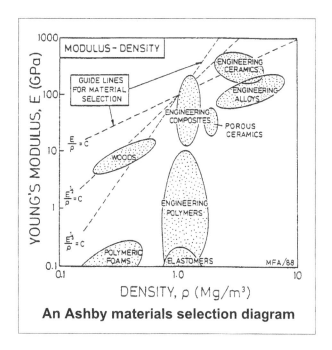

An Ashby materials selection diagram

Why is Most Military Hardware Metallic?

There is an increasing use of polymeric driving bands for projectiles instead of copper based alloys. Personal body armours or 'flak jackets' are made in woven aramid fibres (Kevlar), and sometimes used with ceramic tile inserts. The soldier's 'tin helmet' is now made in a composite material (aramid fibres in an epoxy resin matrix) instead of Hadfield 13%Mn steel. But these examples are rare. **With very few exceptions major equipments, ammunition components, and vehicle armours are made principally of metals.** Yet in the civil sector the rate of substitution to non-metals is ever increasing. The question of why this is so is not easy to answer, and each individual example has different detailed reasons, but in general, the main reason for this is the **superior toughness** of (many) metals compared with polymers and ceramics.

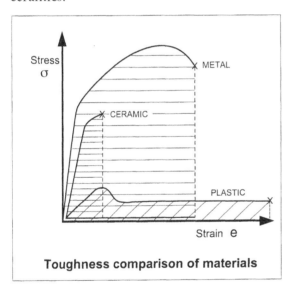

Toughness comparison of materials

The area under the tensile curve is the **energy to fail** per unit volume - highest for the metal with its good combination of strength and ductility. Ceramics are brittle (the curve drawn here with low *E* for clarity) and plastics are of low strength.

A ceramic/plastic composite can show higher toughness than either constituent alone, but joining is a problem (although the rapid development of adhesives technology is encouraging), costs are often high, and fabrication in large sections is as yet rare.

The excellent **fracture toughness** and ultra-high strength combination of the best metallic alloys shows up very well in the Ashby diagram on page 92. Stable and predictable **fatigue behaviour** (over many loading cycles) coupled with fatigue damage repairability are also important considerations, and the best metallic alloys perform very well in these areas. It is obvious that military equipment is roughly handled, requiring it be rugged and not in any way delicate. Contrary to popular belief about military spending, **cost** is an important factor and the Ashby diagram on page 93 shows how well steels perform on a high tensile strength for low cost per unit volume basis. For land-based equipment 'strength to weight ratio at any price' is usually not so critical as it is for an aircraft - when the strength-density Ashby diagram on page 90 would then be more important. Often overlooked is the fact that **Young's modulus** is constant for any particular alloy series regardless of strength, and **Poisson's ratio** (elastic lateral strain over longitudinal strain) is constant at 0.3 for any metal. Both of these parameters vary considerably in non-metals, and even in successively stronger generations of the same material.

The interesting area of **performance at high strain rate** is obviously important in many military applications, and is dealt with in detail in chapter 12. Young's modulus for metals is insensitive to strain rate, which is by no means always true for non-metals. High strain rate **adiabatic heating** is inevitable in ammunition components and armours, and metals with their high thermal conductivity can cope with it much better than polymers or polymeric composites.

There are currently several research programmes aimed (military pun intended!) towards making major equipments in carbon fibre reinforced polymerics **CFRP**. A BR 90 type military bridge made in this material would weigh about 6 tonnes for a 32 metre span, half the weight of the current aluminium alloy, but at twice the price. One CFRP problem to be overcome is that of fracture toughness - the critical defect size **cds** for catastrophic brittle fracture at the 'yield' stress (the 'yield before break' criterion) is around 1mm for a buried defect '$2a$'. This gives rise to concern over barely visible impact damage **BVID** since delicate handling is impossible, and fragmentation and blast damage is likely in battle. Critical defect size for the aluminium alloy is a much more comfortable 9 mm. At least one attempt to make a CFRP ultra-lightweight field howitzer trail leg was shelved in favour of a titanium alloy contingency design. However, there is little doubt that a bulk structure in CFRP, or a CFRP-metal hybrid, will appear in military service before too long.

A common requirement of ammunition components is that they have enough **strength** to survive the stresses of launch, and yet enough **ductility** and **toughness** to avoid brittle shatter on impact at the target. These conflicting property requirements are usually more easily met by metals than by ceramics, plastics, or composites. A good illustration here is the anti-tank **long rod penetrator** dealt with more fully later. It is about 500 mm long and strikes the target at 1500 m s^{-1} or so, close on Mach 5. Its kinetic energy is 11.5 MJ, equivalent to four carriages of a '125' train travelling at 125 mph, and all that energy slams into the target tank delivered on only 25 mm diameter. It is asking a lot of both the ammunition and the armour to not fracture, particularly difficult at this high a strain rate of about 3.10^3 s^{-1}.

2 Brass and Steel Cartridge Cases

INTRODUCTION TO CASED AMMUNITION

The most common type of gun ammunition is the **fixed round** - as sketched below:

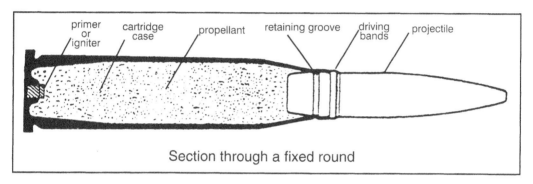

Section through a fixed round

The projectile is fixed into the cartridge case, usually by crimping and often assisted by a retaining (or canneluring) groove. The cartridge case contains the propellant explosive, which is ignited by the primer in the base of the case. The primer may be initiated electrically or by percussion.

In the **gun chamber**, and secured at its rear by the breech block, the cartridge case acts as the combustion chamber for the propellant - as sketched below:

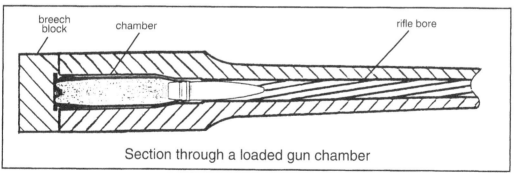

Section through a loaded gun chamber

As the combustion pressure builds, the projectile begins to move and the case mouth expands to give a gas seal with the chamber - called **obturation**. The **driving band** on the projectile body, usually made in soft copper or gilding metal (90%Cu-10%Zn alloy by weight), engraves with the rifling to form the projectile gas seal as it travels up the barrel. The rifling causes the projectile to spin, and **spin stabilisation** in flight prevents tumbling and improves round-to-round consistency. The **calibre** of the ammunition (eg 105 mm) relates to the bore size of the gun barrel, and for a rifled barrel this is the minor diameter - the internal diameter between the rifling 'lands'.

CARTRIDGE CASE FUNCTIONAL REQUIREMENTS AND MANUFACTURE

Large calibre fixed round cartridge cases are usually made in **70/30 brass** (70%Cu - 30%Zn by weight). This alloy is inherently corrosion resistant, not needing paint protection, although sometimes a clear lacquer is applied.

It is perhaps surprising that all calibres of fixed round cartridge cases have an intentional **hardness gradient** along their length - as sketched right, and photographed in Plate 11. The **mouth** is soft and weak to enable crimping onto the projectile body with the minimum of elastic recovery or springback, and also to give early obturation with the gun chamber on firing so minimising burnt propellant gas 'blowback'. Then as well as the increased thickness towards the rear, the **base** has to be hard and strong to withstand the forces of the case **extraction** mechanism after firing.

These requirements dictate the mode of manufacture, which is by the process of **cold deep drawing** from a disc in several stages, with interstage **annealing** (softening) heat treatment at 650°C:

Hardness Vickers	Yield Stress N/mm^2
65	140
95	230
108	250
185	520
210	660

105mm case wall hardness gradient

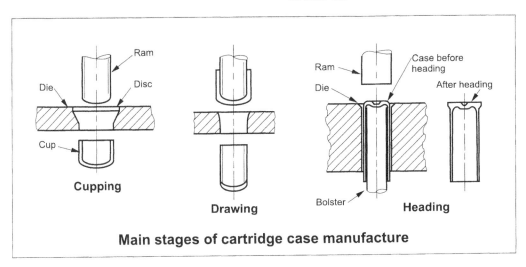

Main stages of cartridge case manufacture

A brass disc, intermediate cup, and finished 105 mm case are photographed in Plate 12. In the latter stages **taper annealing** is employed - the furnace is set with a temperature gradient 250°C at the front to 650°C at the back, and the cases are loaded in mouth first to give the required hardness gradient.

A case could be made in one single step from a disc by **hot deep drawing**, initially heating the disc to 650°C so that it auto-anneals during pressforming, but then the hardness would be 65 Hv all along the case and it would not function properly.

SOME BACKGROUND METALLURGY

Cartridge brass is produced specially for this application. **Residual elements** such as Sn, Pb, Si, Mn and Fe are kept below 0.05 wt%, to give the highest possible ductility for best deep drawability. 70/30 brass is selected so that solid solution strengthening from the Zn is maximised, to give the highest possible strength at the base after cold working. The grain structure is all alpha (α), with an inherently ductile face centred cubic FCC crystal structure, and **annealing twins** abound. These are the 'tramline' features within the grains in Plate 14. If the Zn content is increased to above 33% then its limit of solid solubility in Cu is exceeded, and zinc rich beta (β) grains appear. These have the less ductile body centred cubic BCC crystal structure and contain no annealing twins, as can be seen in the ($\alpha+\beta$) grain structure of 60/40 brass shown in Plate 13. *This alloy, stronger than cartridge brass but not as ductile, is often used for plain bearings where the harder β grains stand microscopically proud helping to retain an oil film.*

These graphs show the results of a popular undergraduate experiment:

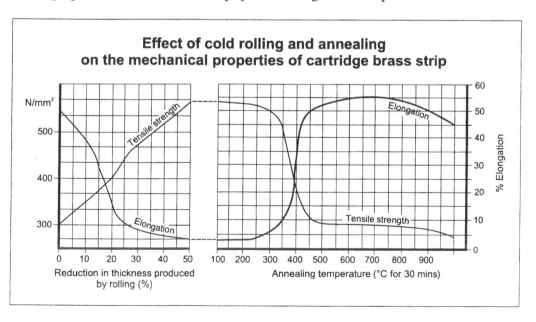

Tensile testing of cartridge brass specimens is done (a) after fully annealed material is put through a rolling mill at room temperature, and (b) after 50% cold rolled specimens have been heat treated in furnaces set at various temperatures.

The first part of the experiment shows tensile strength (UTS) rising and ductility decreasing as the extent of cold work increases - the phenomenon of **work hardening**. The second graph shows that high strength and low ductility work hardened material is unaffected by subsequent heat treatment up to a threshold temperature of about 300°C for this alloy. However, higher heat treatment temperatures cause the strength to fall and ductility to rise until a plateau is reached at above 500°C - this is **annealing** or softening. Then, temperatures higher than 750°C cause a decline in both strength and ductility - the alloy has been 'overcooked'! Commercial full annealing of 70/30 brass is usually done at 650°C for half an hour (nicely on the plateau yet comfortably below the final decline) and this regains the ductility lost by work hardening, taking us full circle back to the start-point of the first part of the experiment.

So, we can plastically deform the alloy (needing ever-increasing force) until about 50% reduction in thickness, when ductility is so low that any further working might cause fracture. But then we can anneal the material ready for further plastic deformation if required. This is the 'what', now for the 'why' :

Plate 14, the microstructure of 70/30 brass after annealing at 650°C, shows a fully equiaxed grain structure containing annealing twins, and the hardness would measure about 65 Hv. **Plate 15** shows the microstructure after 50% cold rolling when the hardness is around 210 Hv. The grains are elongated in the direction of working, appearing less distinct or stained by the etchant - acidified ferric chloride. At higher magnification in **Plate 16** the twins are seen to have been bent during plastic deformation, and the staining is due to the presence of fine lines called **strain lines**. These are the effect of lattice **dislocations** piling up in large numbers in different orientations in each grain. **The yield stress is the stress to move a dislocation**, and during plastic deformation many new dislocations are generated giving a 'traffic jam' effect. It is then more difficult for them to move, and so the yield stress is increased - the explanation of **work hardening**. *Creating barriers against dislocation movement is a classic way of increasing the strength of metals and two other ways of doing this are to (i) reduce the grain size, or (ii) produce very fine precipitate particles within the grains by a heat treatment, and this is 'precipitation hardening' sometimes called 'age hardening'.*

In **Plate 17** the alloy has been cold rolled then annealed at 400°C, about halfway down the UTS drop of the test curve. Some of the grains are elongated (with bent twins and strain lines), but in between there are small equiaxed grains. These new strain-free grains have grown during the heat treatment and this is known as **recrystallisation**. So bulk strength and hardness are reduced, but bulk ductility is increased. The halfway drop position of the UTS curve is arbitrarily defined as the **recrystallisation temperature** (T_R). **Plate 18** shows that when the annealing temperature reaches 500°C the process of recrystallisation has completed, and the strength and ductility changes stabilise. If the annealing temperature is too high, 750°C in **Plate 19**, then **grain growth** occurs (note there is no change of magnification) causing ductility and strength to *both* fall, and of course this is normally undesirable.

So the change in grain structure caused by work hardening can be completely reversed by subsequent heat treatment, and plastic deformation followed by full annealing can be repeated in as many cycles as desired without any adverse effect on the final annealed mechanical properties.

All metals follow this pattern of behaviour, but the temperature scale varies since $T_R \approx 0.3T_M K$, where T_M is the alloy melting point (in degrees Kelvin). **'Cold' working** applies to temperatures below T_R when work hardening occurs, and **'hot' working** is done above T_R when auto-annealing occurs. Two extreme examples are: tungsten (W) with a melting point of 3500°C which is cold worked at 1200°C (when it is white hot!), and lead (Pb) with a melting point of 327°C which when deformed at room temperature is being hot worked!

STRESS CORROSION CRACKING SCC

Although 70/30 brass has good general resistance to corrosion, if stress is combined with certain corrodants then intergranular cracking can occur - as seen right and in Plate 20. This can happen at the unannealed base of a cartridge case due to internal residual stresses, and small cracks here can give catastrophic **premature bursting** in the breech on firing. This problem was encountered in the days of the Raj and known as **season cracking**. It was usual for the ammunition stores to be close to the stables, and in the monsoon season urea from the horses provided the SCC corrodant. This

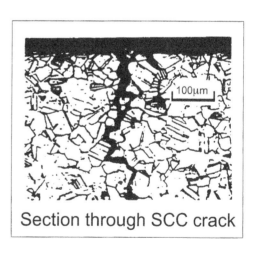

Section through SCC crack

problem was solved by **low temperature annealing** all cartridge cases at 250°C - below T_R. This was sufficient to reduce internal stresses enough to give the cases a shelf-life of about 35 years without fear of SCC, whilst still retaining the high strength required at the base. These days we do not worry about the effect of horses, but sodium chloride is also an SCC agent for this alloy, and the initials **LTA** stamped on the base of a case indicate that the final production heat treatment was at 250°C - and this is particularly common for naval ammunition.

Extraction from the Breech

This diagram is an end-on view of the cartridge case in the breech, exaggerated for clarity.

The **initial chamber clearance** between the case and the steel gun barrel is about 0.7 mm for a 105 mm system. During firing the barrel is pressurised and prevents the case from bursting. After firing - the barrel elastically recovers to its original diameter, but the case is now an **interference fit** requiring considerable force to extract it.

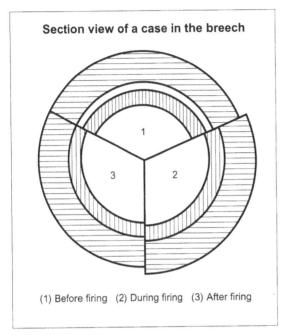

Section view of a case in the breech

(1) Before firing   (2) During firing   (3) After firing

A stress-strain diagram is needed to study the elastic recovery of the case after firing, to decide on the exact initial chamber clearance necessary to **avoid jamming in the breech.** *Note that the stress scale relates to the case wall - the stresses in the gun barrel are far lower since it is much thicker.* In a 105 mm system (here) a mild steel cartridge case will end up with a greater interference strain than a 70/30 brass case. It will jam in the breech (fracture when pulled by the extractor) unless the initial chamber clearance is increased to about 0.9 mm. At this calibre a cartridge case material needs a **minimum yield strength of 500 MPa combined with a Young's modulus (E) less than that of the barrel** to avoid jamming in the breech.

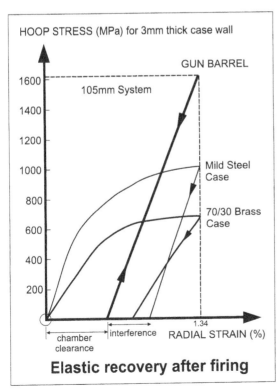

HOOP STRESS (MPa) for 3mm thick case wall

GUN BARREL

105mm System

Mild Steel Case

70/30 Brass Case

chamber clearance   interference   RADIAL STRAIN (%)   1.34

Elastic recovery after firing

For larger calibres, with higher pressure from more propellant, the **envelope** (dotted lines) expands making this problem worse. So recent moves are away from fixed round designs for the largest calibre guns, often using different breech mechanisms, bag charges and **separate loading** projectiles.

For smaller calibres the problem eases as the envelope contracts, and less expensive mild steel cases are used for 40 mm (or less) cannon ammunition - a 25 mm cannon round is seen in Plate 21, painted army green!

Some Possible Alternative Cartridge Case Materials

	MATERIAL	Cost	Corrosion Resistance	Yield Stress (MPa)	E (GPa)
	70/30 Brass	~ £100 (105 mm case)	good	550	110
heavy	Copper	similar to brass	good	250	120
	Cu-2%Be	expensive 6X brass	good	1300	130
light	Aluminium Alloys	cheap	quite good	350	70
	Mild Steel	very cheap £10	poor	600	210
	alloy steel gun barrel			900	210

Copper and aluminium alloys would jam in the breach if used for a fixed round design. The Cu-2%Be alloy is expensive (since Be is radioactive and requires special handling until diluted in the alloy) but its very high yield strength, due to precipitation hardening, would obviate jamming in the breech for a future high charge fixed round if required.

'**Wrapped**' cases - As an alternative to cold deep drawing, brass or mild steel cartridge cases can be made by spiral wrapping cold rolled sheet (like a toilet roll tube) followed by seam welding and welding to the separate base. Mouth annealing would then be done last, although the hardness gradient would not be very gradual. An American made wrapped steel case is seen in Plate 12, and it is 'cleverly' coated in a brass coloured lacquer to help prevent rusting.

3 Steel Shell Bodies - High Explosive Squash Head

The 'high explosive squash head' **HESH** shell is filled with an explosive charge, which is initiated on impact by an inertia fuse fitted in the rear. The detonation on the armour surface transmits a compressive shock wave through the plate thickness. Mode reversal on reflection from the internal free surface then gives a reflected tensile wave which delaminates the armour - shown right and in Plate 22 - and **backspalls** or **'scabs'** detach, acting as **secondary projectiles** inside the vehicle.

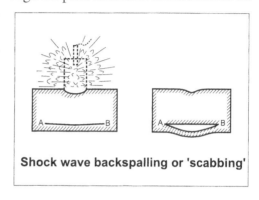

Shock wave backspalling or 'scabbing'

For best 'squash head' performance the nose of the HESH shell has to be ductile and tough enough to give controlled deceleration onto the target (rather than a low energy absorbed brittle fracture) so that the explosive is spread properly into a **'cowpat'** in intimate contact with the target. At the same time the main body has to have sufficient strength to resist **set-back** tensile stresses during launch from the gun tube. These conflicting mechanical property requirements are met in two ways:

The body of the smaller round (fired from the Scorpion light armoured vehicle) is made in a low cost air cooled medium carbon steel, with a separate mild steel nosecap brazed on top.

The larger, heavier, and faster rounds (fired from the Chieftain and Challenger main battle tanks) are one-piece, made in a more expensive low alloy steel (1%NiCrMo) in the quenched and tempered heat treated condition - *see page 94 for steels shorthand notation.*

The resulting tempered martensite microstructure simultaneously gives high strength with good impact toughness.

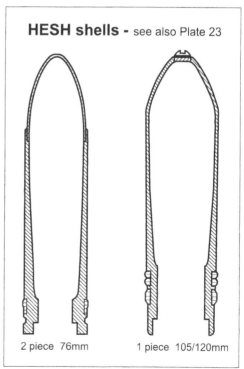

HESH shells - see also Plate 23

2 piece 76mm   1 piece 105/120mm

SOME BACKGROUND FERROUS METALLURGY

Steels are Fe–C alloys and most have a carbon content <0.8%C - *weight % unless otherwise stated*. Unlike many non-ferrous alloys the **cooling rate** from a high temperature heat treatment is very influential on the microstructure, as sketched below and in photomicrographs Plates 24 to 27 :

Air cooled steels - Slow cooling from around 850°C allows two types of grains to form. The white grains are soft ductile iron (**ferrite**) at about 90 Hv local hardness. The dark **pearlite** grains contain the carbon in the form of iron carbide plates, and are stronger and harder at   250 Hv local har dness.

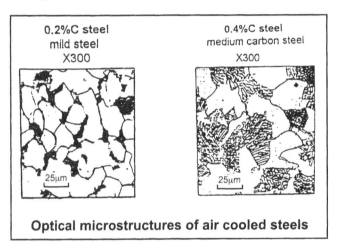

Optical microstructures of air cooled steels

Mild steel with a 75/25 ferrite/pearlite mix has a bulk hardness of about 130Hv. Medium carbon steel has a higher proportion of pearlite, and is stronger with a bulk hardness of 180 Hv , but less ductile. *A high carbon steel with 0.8%C is fully pearlitic, is stronger still with a bulk hardness of 250 Hv, but shows little ductility.*

Water quenched and tempered steels - Quenching a plain carbon or low alloy steel from around 850°C allows insufficient time for iron carbide to form during cooldown. The carbon is retained within **martensite** laths, giving a hard (500+ Hv) brittle steel. Subsequent tempering heat treatment then allows iron carbides to gradually precipitate out as particles (rather than plates) - as seen here, all at magnification X800,   shaded black for clarity.

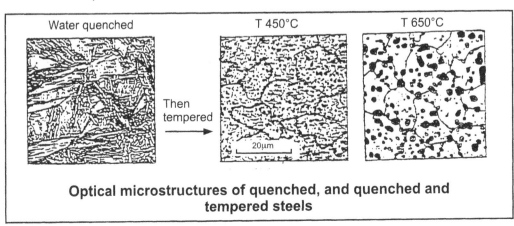

Optical microstructures of quenched, and quenched and tempered steels

After full tempering at 650°C the matrix grains are ferrite and the random dispersion of temper carbides gives a more homogeneous microstructure than that achieved by air cooling, making the steel both stronger and tougher.

For all steels (except austenitic stainless steels) Charpy impact toughness testing at various temperatures shows a **ductile to brittle fracture transition** as test temperature is reduced:

The impact transition temperature T_T is where the fracture is 50/50 ductile/brittle.

The quenched and tempered steel is tougher than the air cooled steel, because its fracture path via the temper carbides is microscopically rougher.

Increasing impact speed encourages brittle fracture in the same way as reducing test temperature does. The faster 105/120 mm HESH body cannot be made with a separate air cooled mild steel nosecap, since its impact transition speed would be exceeded giving brittle fracture at target. The slower 76 mm HESH body could however be made in one-piece quenched and tempered low alloy steel (the alloys Ni, Cr and Mo being present to ensure evenness of quenched properties throughout the full section) - but at greater expense.

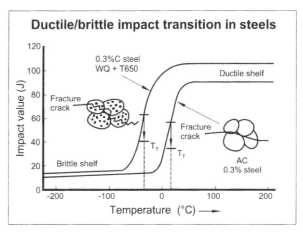

MORE HESH DETAILS

HESH is a way of defeating tank armour without actually penetrating it and regardless of its thickness. The often used analogy is the line of snooker balls – if another ball is run into the back then one ejects from the front, no matter how long the line is – *conservation of momentum*. A scab from a 120 mm round can be up to 30 kg in weight, maybe 600 mm in diameter, and can ricochet around inside the crew compartment with initial speeds of 60 mph!!

The compression and reflected tensile stresses were previously described as **shock waves** – they travel at hypersonic, or supersonic, speed. The **velocity of detonation** (VOD) of a military high explosive is in excess of 8,000 m s^{-1}, and the velocity of sound in steel armour (the same as the elastic wave velocity) is about 5,000 m s^{-1}. The velocity of sound in a metal is given by $(E/\rho)^{1/2}$ where E is its Young's modulus and ρ is its density.

The tensile stress acting on the wall of a shell body during launch is called the **set-back** stress. This arises because the compressive shock wave from the propellant charge acting on the rear of the projectile is reflected back from the nose as a tensile stress. The shell driving band, in forming a gas seal against the gun barrel wall, resists the forward motion of the projectile and this is where the tensile stress is at its highest. So the shell wall thickness is gradually increased towards the driving band.

The production process of **hot rolling** steel (or aluminium alloy) armour plates causes longitudinal alignment of non-metallic inclusions together with microsegregation banding, and these **lamellar weaknesses** assist scab detachment. The important armour material property to resist backspalling is **through-thickness toughness**. The use of 'cleaner' more highly refined alloys, such as electro-slag refined steels, improves anisotropy (directionality) and reduces the likelihood of backspalling. This is dealt with more fully in the 'Steel Armour' chapter later.

Another way of defeating HESH is to have a second layer of **spaced armour** behind, keeping the blunt backspall out.

The high explosive squash head shell is absolutely devastating against **concrete** targets, since although reasonably strong in compression concrete is weak in tension. In the UK it remains popular as the '**second nature**' of ammunition fired from a main battle tank, since it delivers a 'big bang for your buck' against secondary targets. The 'first nature' of ammunition is the long rod kinetic energy penetrator (see later) used against primary targets, ie opposing main battle tanks. Since the advent of multi-layered frontal armours in the 1970's, the long rod penetrator is much more effective than HESH against another main battle tank.

4   Steel Gun Barrels

Rifle, machine gun and cannon barrels are usually made in low alloy steels, such as $1^{1}/_{2}$% CrNiMo or 3%CrMoV, in the tempered martensite condition -*see page 94 for steels shorthand notation*. Here we concentrate on large calibre gun barrels such as the 120 mm fitted to the **Challenger** main battle tank in Plate 55, and the 155 mm fitted to artillery guns like the **SP 70** self-propelled gun in Plate 28 and the **AS 90** artillery system in Plate 29. These are made in **low alloy steel** often **3%NiCrMoV** with 0.3%C (known as 'J' steel), in the tempered martensite heat treated condition for optimum strength and toughness combination.

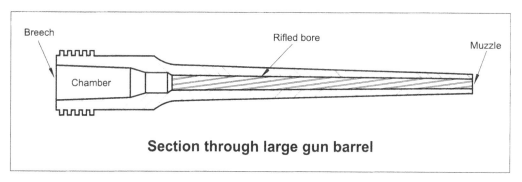

Section through large gun barrel

After forging to shape, the steel is heat treated - oil quenched and fully tempered. This heat treatment gives a **tempered martensite** microstructure - see theory page 22 - to combine a high strength level (YS $\approx$1000 MPa) with good toughness (Charpy at minus 40°C $\approx$ 50 J, $K_{Ic} \approx$ 150 MPa m$^{1/2}$). The latter is needed to minimise the risk of catastrophic brittle failure if a projectile body gets 'stuck up the spout' or if a case 'prematures', or due to fatigue crack propagation - though wearing out before fatigue is more likely.

The Ni content of the steel is sufficient to give full through-hardening to martensite at the thickest section when quenching (up to 150 mm at the breech end) and the Cr, Mo and V carbide formers give high strength after tempering.

SOME OPERATIONAL DETAILS

Direct Fire Tank Guns

The main battle tank gun is the epitome of barrel technology, utilising the highest pressures to give the highest muzzle velocities to the ammunition in order to reduce

target acquisition time to an absolute minimum. Propellant gas peak pressure during firing can be higher than **500 MPa** (5,000 bar) to accelerate a 6 kg kinetic energy round from 0 to 1700 m s^{-1} (Mach 5) along its 7 m length. The long rod penetrator would be fired against an opposing main battle tank at a range of up to 3 km or so, and the 10 kg HESH round fired at lower muzzle velocity against a secondary target at a range of about 5 km.

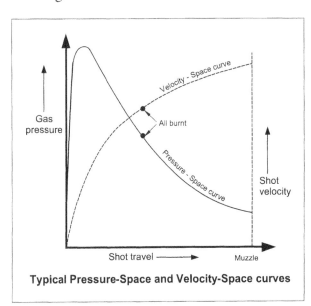

Typical Pressure-Space and Velocity-Space curves

The internal ballistics factors most affecting the design of the barrel are the pressure-space and the velocity-space curves shown left -*after McGuigan RMCS*. As a result of the falling gas pressure, the projectile velocity increases more slowly as it nears the muzzle.

The consistency of a gun is improved by having the position of 'all burnt' as far from the muzzle as possible.

Erosion is very high because of the high temperatures reached on the barrel working surface (>900°C), the chemical aggressiveness of hot propellant gases and the friction from the driving band. The 120 mm L11 gun fitted to late Chieftain tanks (Plate 77) has a rifling depth of about 5 mm, and every time a full charge kinetic energy round is fired an average of 25 microns is worn from the rifling lands - more nearer to the breech and less nearer to the muzzle. This equates to 2 kg of steel dust ejected from the muzzle! Not surprisingly then **this gun is worn out after only 150 full charge rounds**, but well before its fatigue life of about 500 full charge rounds.

In wartime this is not a problem since a main battle tank will only survive an average of 20 minutes in the 'ampitheatre of battle', and it will not fire 150 rounds in that time! But in peacetime the main problem is the price of a replacement barrel and so practice-firings are usually done with three-quarters of the normal propellant charge or even only half-charge.

For the 120 mm guns fitted to the later Challenger main battle tanks the erosion life has been improved to over 500 full charge firings, mainly due to '**hard chromium' plating** on the working surface (more details later). Fatigue life has also been improved to over 2,000 full charge cycles, mainly due to the use of 'cleaner' steels with low non-metallic inclusion content - either electroslag refined steel (more details in Steel Armour chapter), or ladle de-sulphurised 'high-Z' steel. One of these barrels is currently priced at around £75,000.

Even for a **smoothbore** tank gun (often favoured abroad) the wear-life and fatigue life is no better. It is of course preferrable that a barrel wears out before failing by fatigue, whether rifled or smoothbore. The UK slogan is *'Don't be a smoothbore, get rifled !'* and the belief is that a rifled barrel gives greater accuracy than a smoothbore gun, particularly at long range - ie spin stabilised ammunition is more accurate than fin stabilised ammunition. A rifled gun is also needed to fire HESH, which the British prefer to a smaller fin stabilised HE round.

Indirect Fire Artillery Guns

These are sometimes called **howitzers**, though this term falls in and out of favour, and they deliver lower muzzle velocities of 900 m s^{-1} or less. The AS 90 155 mm gun, when at 45° elevation for maximum range, can fire an artillery shell some 25 km (15 miles). Propellant gas peak pressures are a more modest 350 MPa and working surface temperatures are lower. So the barrel survives about **3,000 full charge firing cycles** before wearing out, despite having less deep rifling at 1.5 mm.

Temperature Rise During Firing

Temperature rise during firing causes the strength of the barrel steel to fall as shown right. If the steel temperature reaches 600°C then its yield stress drops to less than half its room temperature value. Fortunately because the barrel wall is thick the outside remains only warm, unless undergoing **'intense rate of fire'** - say 6 rounds per minute for 3 minutes. Of course the tank crew are not at that time most concerned with strength loss, or the wear rate at the barrel working surface even though this increases exponentially with temperature! They are more worried about the loss of accuracy resulting from heat bending of the barrel.

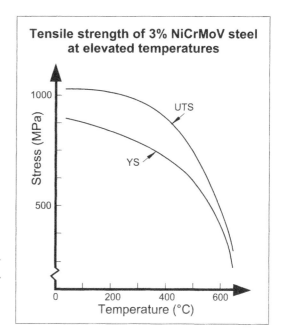

A canvas **thermal sleeve** is fitted along most of the length of a main battle tank gun barrel to counteract the following: Sunlight on top of the barrel causes it to expand and measurably bend downwards or **droop** (affecting accuracy of aim). Rain will cause it to bend upwards, and wind will cause sideways bending. Also, during cooldown after firing, convection inside the barrel causes the top to get hotter than the bottom, giving droop. Of course the thermal sleeve must not be too efficient a heat insulator or barrel wear *will* be seriously aggravated, and there is another problem. If the breech chamber gets too hot then premature initiation of the next round (called 'cook-off ') might occur.

The Muzzle Brake

Plate 30 shows the cast steel muzzle brake fitted to SP 70, which weighs about 100 kg. Its function is to reduce **barrel recoil** after firing, in order to reduce the inboard recoil mechanism mass and to minimise 'ready for next shot' time. The exhaust gases create forward thrust on the muzzle brake baffle dishes, and again erosion wear can be a problem. Some muzzle brakes are made in forged steel segments, electron beam welded together.

A main battle tank gun is not normally fitted with a muzzle brake, since the discarding sabots from the anti-tank long rod penetrator round (more about these in the next chapter) would foul the sides. This round is not fired from artillery guns, so they usually are fitted with muzzle brakes. The American M107 175mm self-propelled gun shown in Plate 31 (now obsolete) is one exception that proves the rule!

PRODUCTION

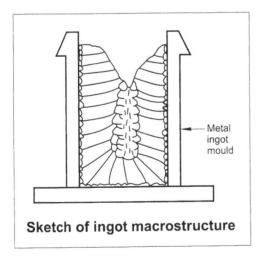

— Metal ingot mould

Sketch of ingot macrostructure

The 105 mm L11 tank gun barrel starts life as a 7 tonne steel ingot. After top and bottom discards of 'pipe' and unsound material this reduces to about 4.5 tonnes, ready for hot **'hollow forging'**. Normally an ingot would be pierced and hollow forged over a **mandrel**. But for large gun barrels (perhaps surprisingly) it is common to firstly bore out the ingot centre to remove the worst of the central inhomogeneities - central porosity, 'V' alloy segregates and non-metallic inclusions. This also helps to maintain straightness during later stages. Hollow forging start temperature is about 1200°C with a finish temperature of 900°C, and its size necessitates three intermediate furnace re-heats. After a **Grain refining** 'soak' at 900°C the forging is slowly cooled to 300°C and held there for several hours - to allow 'outgassing' of hydrogen and the transformation of any unstable austenite (to ferrite and cementite). A final soak at 650°C tempers any martensite that may have inadvertently formed, followed by slow cooling to ambient.

Rough machining is then followed by **final heat treatment** - vertical quenching into oil from 860°C, and tempering at 630°C. Final machining of the inside and outside diameters removes any de-carburised layers, and the barrel is ready for **autofrettage**. The barrel is then rifled by **broaching** and finally **proof-fired**, using a special high-charge round which develops a chamber pressure some 30% greater than a normal full charge. The finished barrel weighs about 1.5 tonnes.

Autofrettage

This process originated in France in the early 1900's and is almost exclusively used for commercial gas cylinders and gun barrels. The gun tube is internally pressurised, either hydraulically after end-sealing or by ramming a swage through it. The resulting few percent

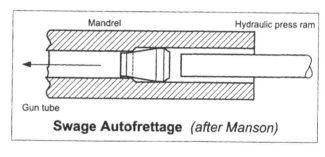

Swage Autofrettage (after Manson)

plastic expansion of the inner diameter is constrained by the outer metal, which afterwards contracts down putting the **working surface into residual compression**, and leaving the outer layers in residual tension. *These residual stresses are considerable, as illustrated when the author hacksawed a half metre length of 120 mm gun from near to the breech for metallurgical investigation. The longitudinal cut was nearly finished when the barrel section sprang open into a 'C' shape with an almighty bang. The area of the fracture was about 600mm² (1square inch) equating to a force of about 100 tonnes!*

Autofrettage results in two main advantages: When the gun is fired the propellant pressure has to overcome the residual compression and the net result is that the complete through-wall stress distribution is more even. Also the at-rest compression tends to close **fatigue cracks**, and they then grow less fast during firing - easily doubling the fatigue life.

WEAR AND EROSION

Over the years a lot of time and effort has gone into research in this critical area of gun barrel technology. As mentioned earlier the temperature of the working surface can easily reach **900°C** during firing, since the propellant gas temperature can be higher than 2200°C when at its hottest (*Lawton, RMCS*). The A_1 temperature of this steel is around 700°C and so some of the tempered martensite on the very surface will transform to austenite with a contraction of about 4% in volume. Then on cooling between firings this austenite will transform to martensite, giving an expansion of about 4% in volume. These localised volume change reversals, combined with macroscopic expansion and contraction during the firing cycles, and the brittleness of the martensite, cause 'craze cracking' of the barrel bore - as seen in Plates 32 and 33. Craze cracking starts during the very first firing cycle whether the gun is rifled or smoothbore. It is a consequence of the ferrite/austenite change temperature (A_1) and is therefore inevitable when using ferritic steel, regardless of the actual grade chosen.

Craze cracking is itself progressive with each shot, but it also allows further sub-surface undermining by the ingress of aggressive chemicals from the propellant. **Plates 34 and 35** show an etched transverse section of a barrel after firing 10 rounds (*from B.Lawton, RMCS*). The first micrograph taken in the optical microscope shows three distinct zones: 'A' is unaffected parent, 'B' is the heat affected zone about 100 µm wide, and 'C' is the chemically affected zone some 5 µm wide. The second micrograph taken in the scanning electron microscope at higher magnification focuses on a crack between zones 'B' and 'C', and also shows an outer thin unetched **white layer**. In-situ energy dispersive X-ray microanalysis of zone 'C' showed decarburisation (from the oxidant in the propellant) and the pick-up of O, S, Si and Ca - all embrittling elements. It is not surprising then that expansion and contraction cycles cause **spalling** of this layer, further aggravated by the scouring action of the projectile driving band. As wear worsens, and especially at the **commencement of rifling** just in front of the breech, '**gas wash**' past the projectile increases causing a reduction of gas pressure and so a lowering of muzzle velocity.

In Plate 33 the barrel bore transverse section is turned to the light, and diamond polishing has revealed the existence of two **fatigue cracks** emanating from the **roots of the rifling**. If this barrel had been fired more often these cracks would have grown further, until reaching a **critical size,** when the next firing would then have caused catastrophic bursting fracture.

This critical defect size **cds** calculates to about 7 mm (worst case surface defect 'a') using **fracture toughness** theory:
The Griffith equation: $K_{Ic} = Y\sigma(\pi a)^{1/2}$
where:

K_{Ic} is fracture toughness
Y is a geometric (compliance) factor
σ is the working stress
a is the critical crack depth (cds)

Taking the working stress as the gun steel yield stress of 1000 MPa, its K_{Ic} value of 150 MPa m$^{1/2}$, and Y as 1, this gives the '**yield before break**' criterion at cds = 7 mm. If a crack is less deep than this value then yielding is bound to occur before catastrophic brittle fracture. If the working stress was half this value (ie 500 MPa) then the cds multiplies by four to become 28 mm. The above 'worst case' calculation would only apply if the gun pressure inadvertently rose to give a barrel hoop stress at the working surface equal to the yield stress of the gun steel - if say a projectile jammed part way along the barrel.

Some Possible Anti-Erosion Measures

It has long been known that **silicone additives** in the propellant (amongst others) can significantly reduce gun barrel wear by providing lubrication, but they also reduce propellant pressure unacceptably.

Water cooling jackets were used for **machine gun barrels** in World War II, well suited for thin walled barrels at relatively low pressures and propellant temperatures especially under **sustained rapid rate of fire** conditions. This would not work so well for thicker walled larger barrels, although there has been at least one attempt at drilling longitudinal holes for cooling water - but this is a little precarious due to radial weakening.

For **rifles and machine gun barrels** the bore wear-life can be enhanced by **nitriding**, or by applying a coating such as **stellite** (a CoCrAlY type alloy often put on by 'plasma spraying') or by **vapour deposition** of say CrNb, or by using **ceramic liners**. To date none of these techniques have worked well for large calibre guns at their higher pressures and temperatures, although a recent American experiment with a **tantalum liner** explosively welded to the inside of a smoothbore tank barrel gun was an interesting attempt.

For **large calibre gun barrels** much research and development effort has been concentrated in the area of bore **chromium plating**. In conventional Cr plating of steel, the component is first dipped into a copper salt solution giving a thin electroless deposit of Cu. Then Ni (the true corrosion preventer) is electroplated onto the Cu, followed lastly by a thin electrodeposited 'flash' of attractive Cr. This works well for rust prevention at room temperature, even though the Cr film is itself slightly porous. However, when high temperature cycled the different thermal conductivities and thermal expansion coefficients of the different layers cause buckling and peeling.

'Hard chromium' plating is preferred for gun barrels. In this process a low porosity Cr film is electroplated directly onto the bore, after having first chemically etched the steel surface to enhance keying-on. This is easier to do for smoothbore barrels, but for rifled barrels a special shaped anode is required and a good 'throwing power' electrolyte is needed to ensure proper coating of the sides of rifling lands. This technique has proved very beneficial towards improving erosion resistance during firing, but if the coating does start to peel then accelerated local erosion can occur.

Interestingly, for a smoothbore barrel with a thicker 1 to 2 mm layer of Cr, it is suggested that the steel substrate never reaches the A_1 transformation temperature and so craze cracking is eliminated.

SOME POSSIBLE FUTURE DEVELOPMENTS

Evolutionary strides are continually being made, with cleaner tougher steels, improving autofrettage, and better hard chromium plating techniques. There is always room for lateral thinking, however, and one or two gun barrel techology revolutions are always possible!

It would seem a good idea to reduce weight by using a higher strength alloy steel to enable a thinner barrel wall to be used for the same propellant pressure. Ultra-high strength martensitic steels might seem to fit the bill, such as HY200 (**AF1410**) 10Ni-14Co-2Cr-1Mo with a YS of 1500 MPa and a K_{Ic} of 150 MPa m$^{1/2}$, or **maraging steel 1700** 18Ni-8Co-5Mo-0.4Ti with a YS of 1700 MPa combined with a K_{Ic} of 130 MPa m$^{1/2}$. Both of these steels, however, are expensive (about 10 times the price of 3%NiCrMoV mainly due to their Co content), they would still suffer from craze cracking, and their fatigue stress would be no higher.

It would seem an excellent idea to eliminate craze cracking by using an ultra-high strength **austenitic** steel (precipitation hardened 'PH'), but as yet their strength levels are not nearly as high as the martensitic steels above.

Liquid Propellants

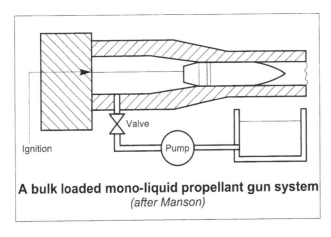

A bulk loaded mono-liquid propellant gun system
(after Manson)

The possible use of liquid propellants LP has been investigated since the end of World War II - either a mono-liquid system as seen left, or a bi-propellant design in which two separate liquids are mixed in the chamber. The latter has the advantage that the two component liquids are only explosive when mixed together. There are several possible benefits of liquid propellant guns including a more rapid rate of fire, and a more gentle pressure-space curve for the same muzzle velocity - which would better suit the so called '**smart shells**' with their more delicate fuse mechanisms.

Electromagnetic Guns

Another idea under investigation over the last few decades is the **EM gun** (or **rail gun**), using electromagnetic energy as a propellant instead of chemical energy. This is discharged into rails, which form the 'barrel' of the gun, giving a powerful magnetic field to then act on an armature at the rear of the projectile thrusting it forward. The rail gun concept would certainly eliminate all of the pressure tube problems, but there would be one or two new ones to solve including **arcing** between the rail and the projectile - back to the metallurgist again to help with that one!

Electro-Thermal Guns

Experimental electro-thermal ET guns use electrical energy to augment thermal efficiency. The simplest type uses a plasma discharge to heat a working medium such as water, which then vapourises to pressurise the projectile. Varying the electrical parameters allows the pressure-space curve to be controlled. However, this design would require considerable electrical energy and a more promising concept is the hybrid electro-thermal-chemical ETC gun sketched here:

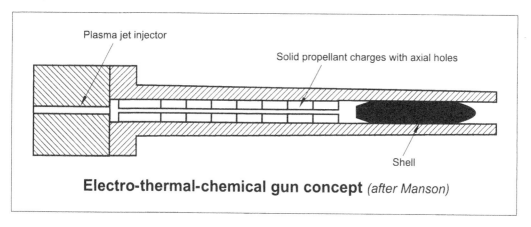

Electro-thermal-chemical gun concept *(after Manson)*

The electrically generated plasma is used to initiate hollow cylinders of solid propellant, and varying the plasma discharge length controls the propellant burn characteristics.

Composite Gun Barrels

An idea with considerable promise is to reinforce a thin steel gun tube with outer layers of say carbon fibre reinforced polymeric CFRP. This could give substantial weight savings whilst still using a steel liner as the working surface. This is not exactly a new notion - Plate 36 shows a 19th century cast iron cannon with peripheral steel wire reinforcing, and it fractured!

5 Heavy Metal Kinetic Energy Penetrators

KE penetrators fired from guns require as high a kinetic energy ($\frac{1}{2}mv^2$) as possible, to literally maximise their impact at the target. Great efforts are made to increase their velocity (v) since this term is squared, but it is also well worthwhile to have a high mass (m) and so 'heavy' metals are used.

ALLOY	Specific Gravity	Density (kg m⁻³)
Steel (Fe)	7.9	7,900
Lead (Pb)	11.3	11,300
Tungsten Carbide (WC) with Co binder	14.0	14,000
Tungsten (W) with 10%NiFe binder	17.0	17,000
Depleted Uranium (DU)	19.0	19,000

They are also designed with as high a length to diameter ratio as possible to give a high value of **'energy density'** - kinetic energy divided by coss-sectional area, usually in J mm⁻² units. This has to be in a compromise with the mechanical properties though, since strength is required during launch and impact toughness is needed to avoid brittle shatter at the target, together with resistance to bending when the target is sloped.

Small arms bullets are made in lead alloys, contained in copper alloy envelopes or jackets for corrosion protection on the shelf. But here we concentrate on **anti-tank KE penetrators**:

The first anti-tank KE penetrators (World War I) were made of solid **steel** at up to 40 mm diameter, but as armour thickness and gun calibre increased the armour piercing discarding sabot round **APDS** was developed. A heavy metal sub-calibre shot is assembled inside a sacrificial segmented sabot, which discards soon after exiting from the gun barrel. The **discarding sabot** principle allows a longer thinner penetrator of higher density to be used without increasing the amount of propellant needed. The **parasitic mass** of the sabot segments is reduced to a minimum by using a lightweight high strength material, usually aerospace aluminium alloy type 7075 - and there have been several experimental sabot designs with other lightweight materials such as magnesium alloys, carbon fibre composites and metal matrix composites. *The word 'sabot' comes from the French for 'hollow wooden shoe' - another military technology idea from across the Channel!*

ARMOUR PIERCING DISCARDING SABOT PENETRATORS

Armour piercing discarding sabot penetrators (**APDS**) with length to diameter ratios of about 4:1 were introduced in the Second World War, and used until the advent of multi-layered 'complex' armours in the late 1970's. Plate 37 shows a sectioned 105 mm round, also sketched here:

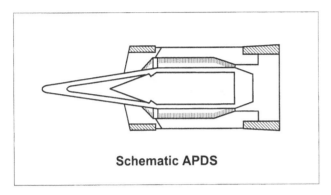

Schematic APDS

With **spin stabilisation** the driving bands were designed to break centripetally shortly after muzzle exit, allowing the sabot segments to discard. The steel **ballistic cap** provided initial armour indentation and reduced the risk of penetrator break-up on initial impact.

A 105 mm APDS penetrator strikes the target with the same kinetic energy as a 10 tonne truck travelling at 60 mph.

The **heavy metal penetrator** (core) was made from tungsten carbide (**WC**) with 5 to 15% by weight of cobalt (**Co**) binder, in much the same way as machine tool tips: WC and Co powders are intimately mixed in a ball mill, then pressure compacted into a 'green'cylinder. This has sufficient strength to enable transfer to a furnace for **liquid phase sintering** at about 1500°C. The melting temperatures of WC and Co are 2900°C and 1495°C respectively, and so the Co binder phase liquates feeding the voids to give full density. The optical microstructure is similar in appearance to that of the tungsten alloy in Plate 41 - though of course the particles are WC and the binder Co. In this form the WC alloy has a hardness of 1500 Hv with poor ductility, and brittle shattering will occur if either the speed of impact or the length:diameter ratio is increased.

Note that in conventional sintering melting does not occur, the particles merely coagulating in the solid state leaving high porosity - and in some engineering applications this is desirable to aid oil lubrication.

ARMOUR PIERCING FIN STABILISED DISCARDING SABOT PENETRATORS

Armour piercing fin stabilised discarding sabot penetrators (**APFSDS** or **'long rod' penetrators**) with length to diameter ratios now as high as 20:1 were introduced in the late 1970's, about the same time as multi-layered 'complex' armours were developed (more on this later in chapter 8). Plate 38 shows an assembled 120 mm round, and plate 39 shows one with the sabots removed to reveal the penetrator core. A section through the complete round is sketched here:

Slipping driving bands are used (currently made in nylon 6,6) to allow **fin stabilisation**, despite the round being fired from a rifled gun barrel. There is some residual spin and this helps with centripetal breaking of the driving bands shortly after exit from the muzzle - seen in a rare photograph, Plate 40.

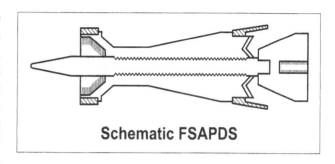

Schematic FSAPDS

A 120mm long rod penetrator strikes the target with the same kinetic energy as half a '125' train (4 carriages) travelling at 125 mph - and with all that energy concentrated on a 25 mm diameter spike!

There is a steel **ballistic cap** (similar to APDS and for the same reasons) and an extruded aluminium alloy **tailfin**. The three 120° sabot segments are extruded in aluminium alloy type 7075, and key on to the penetrator via a 7°/45° 'sawtooth' **buttress thread** - the thrust-face being near vertical. The use of a threadform interface is to maximise the area of metal resisting the shearing stresses during launch. These are considerable since most of the propellant pressure acts on the much larger cross-sectional area of the sabots, which then have to drag the much heavier penetrator up the gun barrel.

The **long rod penetrator** is made from either tungsten alloy W-10%NiFe or depleted uranium:

The **tungsten alloy W-10%NiFe** long rod is made from W (melting point 3400°C) and NiFe (melting temperature 1480°C) powders, **liquid phase sintered** at about 1500°C. A typical final microstructure is seen in Plate 41, the NiFe binder phase etching up black. The larger tungsten particles are around 50 microns in diameter, but smaller ones are necessarily present to provide some 'infill'. The correct mix of W particle sizes (histogram 'cut') is important to give optimum mechanical properties with as high a bulk density as possible. The binder phase should ideally coat every single tungsten particle, and also not form large patches. Final *static* mechanical properties of this alloy are approximately 620 MPa YS, 900 MPa UTS, 15% El, 300 Hv.

The **depleted uranium DU** long rod is made in depleted U-0.75%Ti alloy (nuclear power station 'spent' waste). This alloy melts at about 1130°C and so powder processing is not necessary. The **wrought** alloy rod is heat treated to give **precipitation hardening** and a typical final microstructure is seen in Plate 42. Final *static* mechanical properties of this alloy are approximately 700 MPa YS, 1200 MPa UTS, 7% El, 300 Hv. A similar but lower strength alloy U-2%Mo was used for ship mounted 'Phalanx' anti-missile penetrators until 1993, now superseded by tungsten alloy.

Tungsten alloy versus depleted uranium - The relative technical merits of W and DU for long rod penetrators continue to be discussed by ammunition designers. One advantage of DU is its **pyrophoricity** (oxidising in air) giving a flash on strike and enhancing **'behind-armour' effects**. One advantage of W alloy is its greater tensile **stiffness** - Young's Modulus (E) being 300 GPa compared to 170 GPa for DU - giving reduced deflection under the same stress. However, depleted uranium is currently out of favour, mainly because of 'green' political arguments.

The current 120 mm tungsten alloy long rod penetrator has a length:diameter ratio of 20:1 for high **energy density** at target - being 500 mm long X 25 mm diameter and weighing about 4.5 kg. Strength, ductility, and toughness have all to be in a compromise to survive the stresses of launching (with muzzle velocities of about 1700 m s^{-1}) and yet not shatter at the target. This is more difficult for high length:diameter ratios and if complex laminated targets are to be penetrated.

LONG ROD PENETRATORS AGAINST SPACED TARGETS

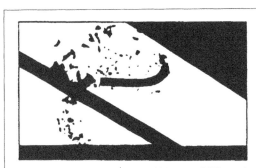

Nose distortion in standard W-10%NiFe rod   Little distortion in swaged W-10%NiFe rod

These flash X-radiographs (*P.Jones, DERA Fort Halstead, UK*) show quarter scale tungsten alloy long rods, flying left to right, penetrating an oblique spaced steel target. This very useful ballistic diagnostic technique uses an X-ray flash of only 10 nanoseconds duration in order to 'freeze' the penetration event.

Neither alloy was brittle enough to break up after penetrating the first plate, but the standard rod has bent and is effectively bluntened for subsequent penetration of the main hull below. The work hardened **'swaged'** rod was trong enough to resist bending distortion, and should therefore penetrate the thicker main hull more easily. Swaging is done on a 'rotary forging' machine – a rolling mill with small outer planetary rolls to give hammering. This is carried out below the recrystallisation temperature, giving work hardening. Note that during work hardening yield strength is increased and ductility is decreased, but tensile stiffness (Young's modulus) is not significantly altered.

HYDRODYNAMIC PENETRATION

The first 10% or so of penetration by a long rod is '**hydrodynamic**' in nature, as illustrated by the series of flash X-radiographs in Plate 43. The use of a copper rod penetrating an aluminium alloy target has enabled differentiation of the two metals as two shades of grey. At very high impact speeds the penetrator tip can be seen to form a 'mushroom head' shape, such that the target hole diameter is larger than the the penetrator diameter. The penetrator material and the target crater both **flow as if they were fluids**. The *dynamic* compressive yield stress of the target is exceeded by a factor of at least 1000 times, such that only the densities of the target and penetrator are important. On slowing down, the hole diameter matches the diameter of the penetrator as the event goes sub-hydrodynamic, and then the *relative* mechanical properties of the two materials do become important.

For a faster moving **shaped charge penetrator**, nearly all of the penetration is hydrodynamic and this phenomenon is described in more detail in the next chapter.

6 Copper Shaped Charge Penetrators

CONICAL SHAPED CHARGE LINERS

The most common **shaped charge** warhead has a copper cone liner fitted at the base of a case containing an explosive charge, as sketched right. When initiated (from 'x') the detonation shock wave emanates spherically, causing the cone to collapse and squirting out a thin high speed **jet** of copper. The jet is capable of penetrating about 9 charge diameters (CD) deep into steel armour. Plate 44 shows **LAW 80** (the modern version of the bazooka) which launches a shaped charge warhead of about 100 mm CD from the shoulder of an infantryman, and will penetrate the frontal armour of a main battle tank. The hole made in a stack of 25 mm thick mild steel target plates is seen in Plate 45, and a selection of shaped charge conical liners is in Plate 46.

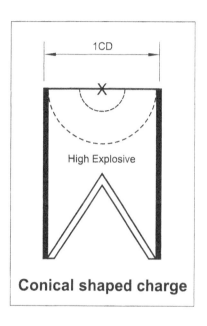

Conical shaped charge

Cone Collapse

The detonation shock wave collapses the cone progressively (Plate 47), giving the characteristic 'sword scabbard' effect. Material flows in a **hydrodynamic** manner towards the centreline, then splits into two streams - one flowing forward as the **jet**, and one flowing relatively rearward to become the **slug**.

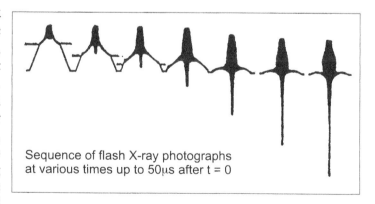

Sequence of flash X-ray photographs at various times up to 50μs after t = 0

Surprisingly perhaps, **jet tip velocity can be as high as 10 km s⁻¹** (Mach 30!), with the jet tail moving at 2-5 km s⁻¹, and the slug at 1-3 km s⁻¹. So jet stretching occurs at a very high strain rate - around $1.10^5 s^{-1}$ - requiring the cone material to have excellent **dynamic ductility** and at temperatures of up to around 450°C.

The **slug**, containing some 80% of the cone mass, follows the jet tail and usually lodges about halfway down the penetration hole playing no part in the penetration deepening process. If warhead build-precision concentricity is poor the slug may not go down the hole, creating instead a shallow impact crater on the target face near to the hole.

The jet tip can move faster than the **velocity of detonation** VOD of the explosive, which is typically around 8500 m s⁻¹, because of **Mach stem intensification**. Detonation shock waves reflected back from the case wall can create a Mach stem in the central region already shocked into higher pressure, and this then moves faster than the primary shock wave.

A cutaway of an experimental 120 mm tank launched shaped charge warhead is seen in Plate 48. A piezo-crystal at the front crushes on impact, sending a signal along an insulated wire to the initiator at the rear of the explosive. The built-in **standoff** tube at the front (about 2 CD long) allows the jet tip to form and reach full speed before meeting the target.

Shaped charge is often called 'High Explosive Anti-tank Warhead', and the acronym **HEAT** *is misleading since the jet does not burn its way through!*

Target Penetration

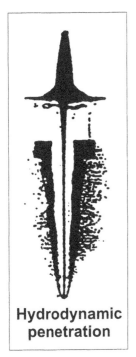

Hydrodynamic penetration

Target penetration is by **hydrodynamic flow**, as seen in the flash X-radiograph of Plate 49 and reproduced left. Hypervelocity hydrodynamic impact (unlike lower speed KE penetration) results in a 'mushroom head' tip, and the hole diameter is larger than the penetrator diameter.

The **dynamic compressive yield stress** of the target is exceeded by a factor of at least 1000X, such that only the densities of the target and jet media are important. Both **flow as if they were fluids** and the penetration event can be modelled quite accurately using Bernoulli fluid flow equations (more later).

However, **X-ray diffraction** shows the jet to be solid metal and not molten. Also best estimates of jet temperature by incandescence colour (*Jamet*) suggest an average of about 450°C, and the melting point of copper (the usual liner material) at atmospheric pressure is 1083°C. So:

The jet appears to behave like a fluid, and yet it is known to be a solid.

One recent theory is that the jet has a molten core, but with a solid outer sheath (*I. Cullis, DERA*) - and this would help explain the conundrum.

Some Facts and Figures

It is difficult to think of many terrestrial events as fast as a jet tip, and some other shaped charge facts are just as sobering:

The jet tip reaches **10km s⁻¹** (Mach 30) some 40 µs after detonation, giving a cone tip acceleration of about **25 million g**. At this acceleration the tip would reach the speed of light (were this possible) in around 1.5 seconds, but of course it reaches a terminal velocity after only 40 millionths of a second. On meeting a target the pressure then developed between the jet tip and the forming crater is about **10 Mbar** (10 million atmospheres), several times the highest pressure predicted in the Earth's core.

Shaped charge is truly an extraordinary phenomenon, 'off the scale' of 'normal' physics, and its fundamental mechanism is not fully understood.

The Penetration Equation

At constant standoff the effect of liner and target densities can be predicted using the Hill, Mott and Pack [1944] **hydrodynamic penetration equation:**

where P is penetration, L is jet length, ρ_j and ρ_t are the densities of the jet and target respectively, and λ is a warhead constant (1 to 2) associated with jet lengthening.

$$P = L\sqrt{\frac{\lambda\rho_j}{\rho_t}}$$

The equation is derived assuming Bernoulli fluid flow behaviour - conservation of mass, energy and momentum is applied either side of the stagnation point, where material flowing in equals material flowing out. It works well for a wide range of liners and targets, despite its incorrect simplifying assumption that there is no velocity gradient along the jet.

It is clear that target penetration P is improved if jet density is increased, but only if jet length remains high. A good copper jet, inherently ductile due to its FCC crystal structure, will be 8 CD long in air before its starts to particulate. For **ductile metal liners** where L is fairly similar, the equation correctly predicts penetration into steel in cone density order - copper, mild steel and aluminium having densities of 8.9, 7.9, and 2.7 (specific gravity units) respectively. *Jet density is the same as the cone density for metals, but for* **polymeric** *cones flash X-ray contrast shows the jet to be less dense than the solid polymer.*

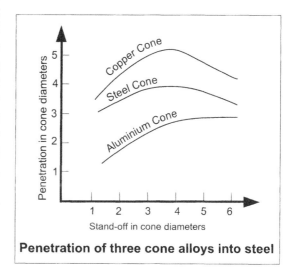

Penetration of three cone alloys into steel

Liner materials research (more later) is thus often driven towards **high density metals**, but many of these are not FCC and are much less ductile than copper, giving **lower L values** and negating their higher density.

Some candidate pure metals are:

METAL	Cu	Pt	W	Au	DU	Ta	Pb	Ag
density (sg)	8.9	21.4	19.3	19.3	18.9	16.6	11.3	10.5
$\sqrt{\rho_m / \rho_{Cu}}$	1	1.55	1.47	1.47	1.46	1.37	1.13	1.09
crystal lattice	FCC	FCC	BCC	FCC	HCP	BCC	FCC	FCC
mpt (°C)	1083	1772	3410	1066	1132	2996	327	962

In theory then a **gold** cone (for instance) would be capable of penetrating 47% deeper than copper into the same target, if the jet was no shorter - and its FCC crystal structure would give a reason to be optimistic about this. However, gold is usually regarded as being too expensive!

Copper Cone Manufacture

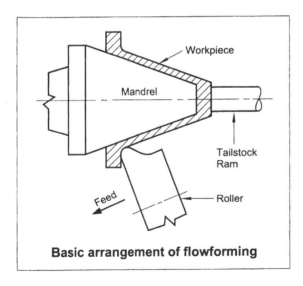

Basic arrangement of flowforming

In the UK copper cones are usually produced by **flowforming**: An annealed copper plate, or blank, is held by the lathe tailstock ram against the mandrel. The roller tool **plastically deforms** the plate over the mandrel at room temperature to achieve the desired cone shape. Note that the cone wall is thinner than the original plate because of the plastic deformation, which also causes **work hardening**.

After machining off excess flash material from the rim, the cone is **annealed** - heat treated at about 500°C for 30 minutes - to remove the work hardening by recrystallisation of the grain structure, returning the cone to the fully softened (most ductile) condition. The aim is to achieve **fully equiaxed copper grains** in the finished cone, at less than 30 microns MLI grain size.

OTHER SHAPED CHARGE LINERS - EFP'S

Wide angle cones and other liner shapes such as **plates** or **dishes** do not jet, but give instead an explosively formed projectile **EFP** – sometimes called a self-forging fragment SFF. The fragment or slug forms by

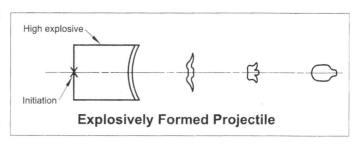

Explosively Formed Projectile

plastic flow and has a velocity of 1-3 km s⁻¹. Target penetration is much less than that of a jet, but hole diameter is larger with more armour backspall. EFP's are less sensitive to standoff than jets, and so can be initiated from several tens of metres away from the target. They are popular for mines and Overhead Top Attack **OTA** warheads, targetting the thinner armour of the belly and the tank-top respectively

SOME LINER MATERIALS RESEARCH

Despite considerable research and development effort on alternatives, **copper** has remained a favourite conical liner material for several decades, and yet **iron** and **tantalum** perform better for EFP liners. Confusion like this is common in the shaped charge field. **Cartridge brass** is more ductile than copper and yet performs less well when tried as a shaped charge cone. **Lead** is an interesting candidate - it is FCC with a higher density than copper, and its low melting point would ensure a molten jet - it is truly 'hydrodynamic', and yet in practice it underperforms by a considerable margin. Because of better ductility, copper cones with a finer **grain size** perform better than those with larger grains, and yet a finer grain size also confers higher strength. **Graphite cones** and even **ceramic cones** with zero ductility have shown decent penetration into steel targets.

Copper is an excellent shaped charge penetrator, but it does not oxidise with any voracity (poor pyrophoricity) and so its **behind armour effect** BAE is limited to backspall with only minor temperature and pressure rises. Research with more pyrophoric alloys such as Zn-Al has shown excellent behind armour effects, but their low density curtails penetration.

Computer modelling using 'hydrocodes' is an important research technique. For best accuracy the models need to encompass a host of strength properties for the liner and the target materials at various strains, strain rates and temperatures - and yet hydrodynamic deformation is supposed to not depend on them. The situation gets very complex when considering the attack of **multilayered armours**, and their advent together with reduced availability of real firing trials has meant increased reliance on mathematical modelling.

Some Other Variables Affecting Penetration Performance

Apart from the liner material and its mechanical properties, there are many factors which affect penetration performance including:

Charge diameter - A conical shaped charge at twice diameter will penetrate the same target twice as deeply (provided it has twice the standoff) even though the explosive velocity of detonation remains the same. This linear scaling is very useful, enabling penetration and standoff to be expressed in terms of charge diameter.

Standoff - Penetration rapidly decreases above about 8 CD standoff for a conical shaped charge. This is caused by 'lateral velocity' of jet particles, such that later particles may not go down the hole, giving widening instead of deepening. A high symmetry precision-built device will thus perform better at higher standoffs.

Cone geometry - Cone angle, wall thickness, and tip radius are all prime variables requiring optimising for any particular liner material. In general as cone angle is increased penetration reduces - the jet gets fatter, heavier, and moves more slowly. The jet is also slowed down if cone wall thickness is increased beyond optimum, and cone tip geometry greatly influences the shape of the jet tip.

Case confinement - The charge case is important since it reflects the primary detonation shock wave back towards the liner. A thinner and less rigid case will cause a reduction of target penetration.

Charge height, explosive type, and initiation method - These also require optimisation.

These variables are often interlinked and so experimental firing trials have to be designed very carefully to avoid drawing misleading conclusions.

Shaped Charge Weapons Systems

As well as various man portable light 'anti-armour weapon' LAW and 'rocket propelled grenade' RPG type anti-armour shaped charge weapons, other military shaped charge devices include:

Anti-tank guided weapons ATGW - Milan, Swingfire, Trigat, Hellfire, Bofors Bill, and Copperhead. Merlin and Strix are mortar launched ATGW's.

Torpedoes - Stingray and Spearfish anti-submarine torpedoes both have shaped charge warheads.

Bomblets – The M42 shaped charge bomblet can be deployed in large numbers from a carrier shell, fired from either the 'multi launch rocket system' MLRS or the 155 mm gun.

Cluster Bombs - Delivered from aircraft pods, the JP233 'runway buster', and the BL755 anti-tank 'top attack' devices both utilise shaped charge.

7   Ferrous Fragmenting Projectiles

The requirement for a casing to deliberately fragment in service must be unique to the military. The high explosive filling is expected to cause the shell to burst in a reasonably predictable manner, giving an optimum number and size of fragments to act as omni-directional secondary projectiles. The **velocity of detonation** of the explosive is about 8000 m s^{-1} and the detonation wave expands at a rate faster than the speed of sound in the shell - around 5000 m s^{-1} if the shell is steel. This **shock wave** will cause unpredictable brittle shattering of the casing ('**brissance**') if the material has insufficient ductility.

CAST IRON MORTAR BOMB BODIES

Plate 50 shows the 81 mm mortar, with the bomb photographed in Plate 51 and also sketched here.

The **81 mm mortar bomb body** has a smooth wall and is made in **cast iron** to assist its fragmentation.

Some fragmentation devices (eg BL755 bomblet casing) have **internally** notched walls to ensure breakup, and their material properties are less important. The old Mill's bomb hand grenade had **external** notching ('pineapple chunks') but this is now thought to have not worked too well.

Cast irons are Fe-C alloys with about 4%C by weight giving free carbon in the microstructure in the form of brittle graphite, resulting in low tensile strength *and* ductility. The optical microstructure of **flake grey cast iron** is seen in Plate 52 - graphite flakes in a ferrite matrix. Tensile UTS is around 230 MPa and ductility is only 2 %El, greatly assisting fragmentation. It is called 'grey' because its fracture surface is less silvery than steel, the colour being dulled by the graphite.

81mm mortar bomb

*Flake grey cast iron is often called **automobile iron**, since it is the most popular material for car cylinder blocks. The graphite flakes assist greatly with machinability by causing the swarf to break up and by lubricating the tool tip - water based lubricant is really only needed to dampen down the fine grey dust. Also in use the graphite attenuates the internal combustion sounds. In comparison, aluminium alloy cylinder blocks and heads are notoriously noisy.*

The optical microstructure of **spheroidal graphite cast iron** ('sg iron') is seen in Plate 53 - graphite nodules in a ferrite matrix. Before casting, a small amount of rare earth metal (REM lanthanum and cerium - Misch metal) is added which alters the surface tension properties between the molten iron and the carbon, causing the graphite to form into spheres. This improves the tensile properties, and sg iron has a typical tensile UTS of 430 MPa with a ductility of 18 %El.

Until recently flake grey iron was used for the **smoke** dispensing mortar bomb where fragmentation pattern is unimportant, and sg iron was used for the **HE** anti-personnel variant. In the latter the blunter and more even dispersion of the brittle graphite phase gives higher numbers of fragments and in a more repeatable pattern round to round. In 1993, with the closure of the Royal Ordnance Factory at Patricroft (Manchester), UK production was rationalised and both rounds are now made in the slightly more expensive sg iron - *spoiling a very nice story relating microstructure to properties!!*

STEEL 155 MM ANTI-PERSONNEL ARTILLERY SHELL BODY

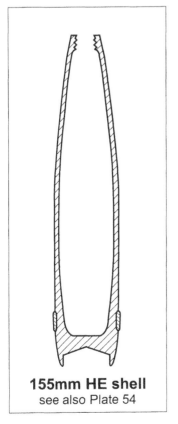

155mm HE shell
see also Plate 54

This much larger and faster thinner walled shell has to have greater tensile strength to resist the much higher **set-back stresses** during launch and steel has to be used. Most steels have much higher strength and ductility than cast iron because their lower %C means the carbon is present as comparatively finely divided iron carbide Fe_3C. So a rather unusual (perhaps surprising) approach is used to reduce the steel ductility and toughness, in order to optimise fragmentation - **temper embrittlement** is deliberately induced:

The shell is made from **2%SiMnCr spring steel** - *see page 94 for steels shorthand notation*. This is a BS970 - 250A58 grade steel (En45A) normally used for automobile road springs - but **modified** by having a high carbon content (around 0.7%C), an increased chromium content (from 0.1% to 0.5%), and a low molybdenum content (0.02%Mo max).

After forging from billet the shell is heat treated - 880°C water quench to martensite then tempered at about 620°C and air cooled, resulting in a typical tensile UTS of 1100 MPa with a ductility of 8 %El and a Charpy impact value of only about 10 J.

For the unmodified spring steel, temper embrittlement is positively avoided by ensuring a molybdenum content of about 0.4% and water quenching instead of air cooling after tempering. These two changes avoid the formation of an embrittling grain boundary Mo_xSi_yC type temper carbide precipitate, and give a Charpy impact value of about 30 J (much more conducive to higher fatigue life of the spring!).

METALLURGICAL QUALITY CONTROL FOR FRAGMENTATION

It would be unrealistic of the end-user to specify an explosive burst test for the steel supplier to use for quality control purposes, although the munitions factory would carry out the occasional field test (almost literally!). It is better to try and relate the desired final performance to everyday mechanical properties such as YS, %El, and Charpy impact toughness, and this is best done at the munition research and development stage.

This approach is by no means unique to military devices, but the problem of fragmentation is particularly difficult. The relationship between fragmentation performance and the common mechanical properties seems particularly complex, and as yet is poorly understood.

8 Steel Armour for Main Battle Tanks and the Milne de Marre Graph

STEEL ARMOUR PLATE

The hull of a main battle tank **MBT** such as **Challenger**, photographed in Plate 55 and sketched below, is fabricated by welding 'rolled homogeneous armour' **RHA** steel plates together.

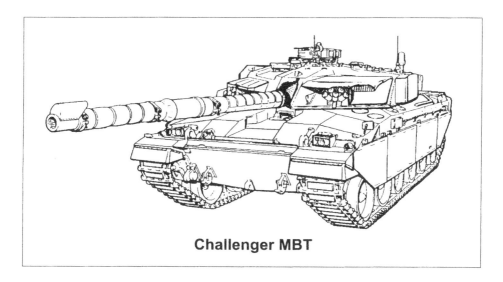

Challenger MBT

Plates **up to 100mm thick** are made in **low alloy steel - 1¹/₂%CrNiMo** BS970 - 709M40 grade (En19), water quenched and fully tempered to the UTS 850 MPa level - *see page 94 for steels shorthand notation. This is common or garden automobile crankshaft steel and it is perhaps surprising that the word armour in 'armour plate' holds no special significance, but don't tell the media people!!*

Plates **over 100mm thick** are made in **low alloy steel - 1¹/₂%NiCrMo** BS970 - 817M40 grade (En24), also water quenched and fully tempered. The extra Ni improves the quench **hardenability** to give the strong and tough tempered martensite microstructure through to the plate centre in these greater thicknesses.

The term **rolled homogeneous armour** refers to these plates being 'hot rolled' as opposed to 'cast', and the steel being of uniform (homogeneous) microstructure as opposed to 'face hardened' :

Cast steel armour (used for the turret and other complex shapes) not being hot rolled has a less refined grain structure with inferior mechanical properties and has to be some 10% thicker for the same ballistic resistance as RHA. Recent more angular turrets are sometimes fabricated by welding RHA plates together rather than casting, to save this extra weight.

Face hardened steel armour was used on World War II German King Tiger tanks. Plate outer faces were flame hardened to give **dual hardness** (as opposed to 'homogeneous' single hardness) - the hard martensite face encouraging **shot shatter**, and the tough core arresting the brittle microcracks. Dual hardness can also be achieved by face **carburising** and Plate 56 shows a through-thickness section after kinetic energy (KE) attack. This approach is currently out of favour except for thin plates on helicopter seats and on warship electronic module boxes.

ARMOUR FAILURE MECHANISMS AGAINST KINETIC ENERGY ATTACK

A long rod penetrator flying in almost horizontally will pierce the thick sloped front **glacis plate** of the MBT at an angle. This **obliquity** effectively thickens the armour for no weight penalty, and causes curving of the penetration tract - see Plate 57. Apart from the possibility of bending fracture of the long rod, this makes analysis of the armour failure mechanism complicated and most studies concentrate on 'normal' (90° angle) attack:

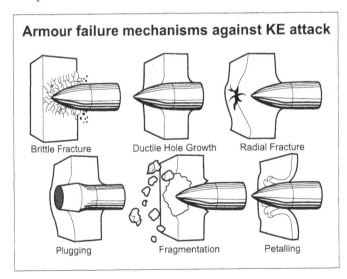

Armour failure mechanisms against KE attack

Brittle Fracture   Ductile Hole Growth   Radial Fracture

Plugging   Fragmentation   Petalling

Petalling occurs if the armour is too thin, bulging then giving rise to star cracks on the inside face which propagate to failure. **Fragmentation** is due to lack of plate through-thickness toughness.

Radial fracture and **brittle fracture** are due to lack of general toughness in the plate. **Plugging** can occur with a blunter and/or softer projectile, or if the armour is susceptible to **adiabatic shear** - more on this in chapter 12. A through-thickness section of a plugging failure in aluminium alloy armour is seen in Plate 58. **Gross cracking** is a rare type of armour failure, shown in Plate 59. In this case the steel plate had been quenched but inadvertently not tempered, and its brittleness is clear.

Ductile hole growth is the preferred armour failure mode - this plate has sufficient toughness to avoid any kind of cracking, and it is therefore capable of absorbing the most penetrator energy.

IMPROVING THE THROUGH-THICKNESS TOUGHNESS OF STEEL

In an armour plate the through-thickness or short transverse direction is the very direction being attacked, and improvements to the toughness in this particular direction should contribute greatly to improved ballistic resistance.

The author now describes the essence of some of his own research carried out with this particular aim in mind - *done jointly with Brian Neal whilst at Aeon Laboratories, Surrey*. This work was on a **3%NiCrMo low alloy steel armour** in the quenched and fully tempered heat treated condition, though the principles would apply to any grade of low alloy steel:

Plate 60 shows the low magnification optical microstructure of the through-thickness section of a high quality air melted thick plate of this material. The appearance of **microsegregation banding** was very clear in this rarely studied direction - alternate lean and rich alloy content bands, remnant dendritic and interdendritic regions from the ingot casting despite extensive hot rolling (thermomechanical reduction) down to finished plate. At higher magnification in Plate 61, a **manganese sulphide MnS non-metallic inclusion** is seen in a dark etching rich alloy band - trapped in the last liquid to solidify in the ingot. The poor performance of short transverse Charpy impact specimens (relative to the longitudinal and long transverse test directions) after testing at room temperature was attributed to the fracture crack being attracted to the dark etching bands and their resident MnS inclusions - Plate 62. Even when fully brittle after impact testing at minus 196°C, the fracture crack was still attracted to the MnS inclusions, as seen in the SEM fractograph of Plate 63.

Further trials on plates rolled from electroslag remelted **ESR** ingots rather than air melted ingots showed a marked improvement in short transverse impact toughness, due to the reduction in MnS inclusion population and to decreased microsegregation banding (Plate 64) resulting from the ESR process. The process is shown in Plate 65 - an air melted billet is remelted under a calcium fluoride containing slag, which removes many of the MnS inclusions. Also the small molten pool size leaves little time for microsegregation to occur during freezing, and a diagram comparing ingot macrostructures is shown in Plate 66.

Ballistic trials showed that electroslag remelted armour plates would not scab, making them **proof against HESH attack**, and their general ballistic resistance was significantly enhanced too. More latterly similar gains (though lower) were found using less expensive **ladle de-sulphurised air melted steels**. HESH can be rendered ineffective by using spaced armour anyway, but the principle of improved ballistic resistance via improved short transverse microstructure and toughness is clearly indicated.

COMPLEX MULTI-LAYERED FRONTAL ARMOUR

The frontal armour of many modern main battle tanks is of multi-layered 'complex' construction rather than thick monolithic steel. The main hull is still RHA steel but other materials are sandwiched between it and outer steel plates, and a good outermost **applique** layer is a mosaic of ceramic tiles (to encourage shot shatter) covered in radar absorbent 'paint' (green!). An alternative applique is **Explosive Reactive Armour** ERA, consisting of bolted on steel boxes containing sheet explosive - Plate 67. The explosive is initiated by an incoming shaped charge jet, but not by small arms kinetic energy KE, and the box roof and floor fly apart consuming much of the jet before it reaches the main armour below.

THE MILNE DE MARRE GRAPH

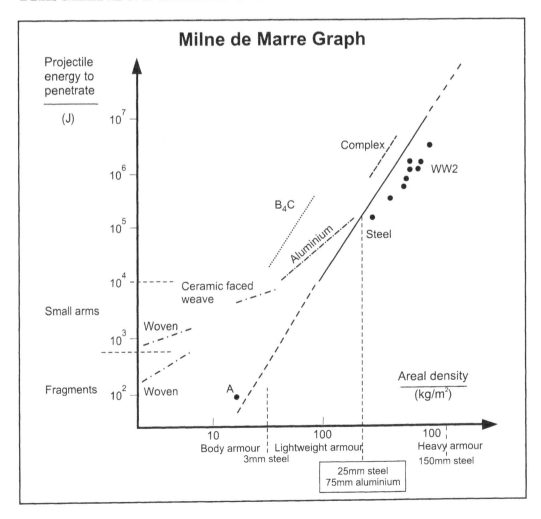

This empirical graph is for KE penetration of various armour materials, plotting projectile energy to penetrate versus thickness of the armour. Note that the scales are logarithmic (to give the straight lines) and that the thickness scale is expressed in terms of **areal density** - the mass of 1 square
metre of the armour. This term is much favoured by armour designers, who are usually more interested in weight per unit area than in actual thickness.

Milne and de Marre started this empirical graph by plotting point '**A**' - the energy of the arrows of **Agincourt** at the thickness of the (steel) chain mail body armour of the day. This gave the first point on the steel armour line, extending up to the dots of the various tanks of World War II.

At plate thicknesses greater than 25mm 'steel equivalent', steel armour gives a better protection/weight ratio against KE attack than does aluminium alloy armour - since the projectile energy required to penetrate it is higher. So steel armour is used for main battle tanks. A *25 mm thick steel plate weighs about the same as a 75 mm thick aluminium plate of the same area - the densities of steel and aluminium being 7,900 kg m⁻³ and 2,700 kg m⁻³ (roughly 3:1).*

At plate thicknesses below this crossover point, aluminium alloy armour is superior to steel by virtue of its lower line-slope. So aluminium alloy armour is advantageous for thinner gauge light armoured vehicles (LAV's) such as armoured personnel carriers (APC's) and mechanised infantry combat vehicles (MICV's).

Modern body armours - 'Flak jackets' are woven from high strength polymeric fibres such as **Kevlar** (an aromatic polyamide), sometimes with ceramic panel inserts, and the lines for these materials sit comfortably higher than the Agincourt 'A' - *they are indeed more protective and lighter than chain mail!* If the lines are extrapolated upwards, the graph also indicates that these armours would be more ballistically efficient than aluminium at up to 9 mm aluminium thickness (12 mm thick Kevlar), and more ballistically efficient than steel at up to 12 mm steel thickness (50 mm thick Kevlar) - assuming the same ballistic integrity can be maintained at these greater lay-up thicknesses.

The protective superiority of **boron carbide** ceramic armour, and of **complex** multi-layered armour is also quantified by the Milne de Marre graph.

9 Aluminium Alloy Armour for Light Armoured Vehicles

Aluminium alloy light armoured vehicles (LAV's) first emerged in the early 1950s, being designed with **air-transportability** and **air-droppability** in mind for rapid deployment. Not only does the Milne de Marre graph show that aluminium alloy plate at less than 75 mm thick gives a better protection/weight ratio than steel, but its greater bulk means that fewer structural stiffeners are needed and this gives further weight savings. However, these vehicles only offer protection against small arms, rifle-fire, and air-burst HE fragments - they are no match for long rod penetrators and shaped charge warheads. In the **mobility-protection-firepower** triumvirate, the accent is very much on the mobility side - these vehicles are supposed to manoeuvre away rapidly from real trouble!

M113 ARMOURED PERSONNEL CARRIER ARMOUR

The American M113 APC (Plate 68) was the first military vehicle to be fabricated from aluminium alloy plate, and weighs around 7500 kg. It was developed in time for the Korean conflict and several thousand are still in service today.

M113 is made in type 5083 alloy **Al-5%Mg** in the **20% cold rolled condition**. This alloy is 'non heat-treatable' (meaning non precipitation-hardenable), and its strength (tensile UTS 390 MPa) is derived from the Mg solid solution strengthening and the 20% cold rolling - giving elongated work hardened grains as shown in the micrographs of Plate 69.

Welding of the plates is by metal inert gas **MIG** - an electric arc technique where the gun shroud feeds argon inert gas over the work to prevent oxidation (right). The consumable electrode filler is the same alloy as the parent metal. The fusion welding heat causes annealing, softening the alloy to UTS 320 MPa, necessitating thicker plate edges to compensate.

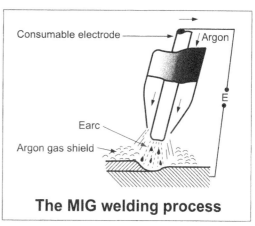

Consumable electrode — Argon

Earc

Argon gas shield

The MIG welding process

SCORPION COMBAT RECONNAISSANCE VEHICLE ARMOUR

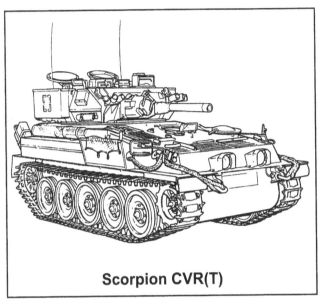

Scorpion CVR(T)

Scorpion combat vehicle reconnaissance (tracked) CVR(T) - Plate 70 and sketched left - was developed after M113 and weighs just under 7000 kg.

Scorpion and its variants are made in type **7039** aluminium alloy **Al-4Zn-2Mg** with high strength derived from a **precipitation hardening heat treatment** (or 'age hardening') - 450°C WQ + age at 90/150°C. This produces ultrafine Zn-Mg precipitates within the grains, shown in the electron micrograph of Plate 71, which block dislocation movement giving strengthening to UTS 475 MPa.

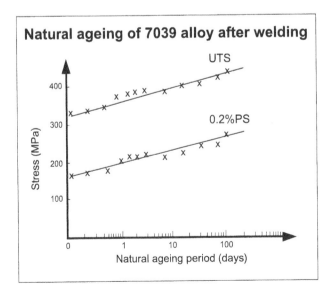

Natural ageing of 7039 alloy after welding

MIG welding (using Al-5%Mg filler rod) re-solutionises the precipitates causing annealing to about UTS 320 MPa, but fortunately '**natural ageing**' at room temperature then slowly allows some re-precipitation. This restores the strength to around UTS 420 MPa after 100 days. *After any repair welding, Scorpion is supposed to be put on light duties for 3 months while it strengthens up!*

Stress corrosion cracking SCC is known to be an occasional possibility in this alloy. A corrodant combined with a stress can give cracking, often via grain boundaries as shown in this low magnification optical micrograph - taken near to a weld.

Exposed plate edges near to a weld joint are sometimes '**buttered**' with Al-5%Mg welding rod (shown in Plate 72). This is done to relieve internal stresses and also prevent possible ingress of a corrodant.

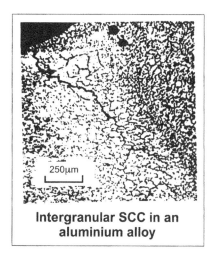

250μm

Intergranular SCC in an aluminium alloy

WARRIOR INFANTRY FIGHTING VEHICLE ARMOUR

The Warrior infantry fighting vehicle IFV - Plate 73 and sketched right - is metallurgically very similar to Scorpion, being made in **Al-4Zn-2Mg** alloys types **7017** and **7018** (both close relatives of 7039). The former is full strength at UTS 485 MPa for ballistic plates, and the latter is 'overaged' to UTS 350 MPa for structural members to give full resistance to SCC.

The turret is fabricated in 'rolled homogeneous armour' steel, and

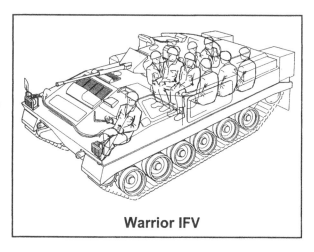

Warrior IFV

the loaded vehicle weighs about 24 tonnes - very similar in weight to the American Bradley IFV, seen in Plate 74 - and very close to the maximum payload of the C-130 Hercules aircraft.

Possible Alternative Armour Materials for Light Vehicles

'**High Hardness**' **Steel** – The ballistic resistance of thin plate 300 Hv 'rolled homogeneous armour' steel can be improved by raising the hardness to about 500 Hv, provided the impact toughness is not seriously lowered. This can be achieved by lowering the tempering temperature down to around 300°C, altering the alloy content slightly, and using 'cleaner' ladle de-sulphurised steel - as detailed in chapter 8. On the Milne de Marre graph this elevates the steel line and displaces the aluminium/steel crossover point to the left. 'High hardness' steel panels are currently available as extra side-armour to **upgrade M113**, and are used for the entire armour of the **Vickers Valkyr** reconnaissance vehicle (sold abroad). The use of this armour material is likely to increase.

GFRP - In 1989 the American FMC Corporation completed the construction of a prototype **glass fibre reinforced polymeric** GFRP hull for an infantry fighting vehicle. They chose strong S-2 type glass fibres in the composite to give best ballistic protection, and manual laying up of the pre-pregs enabled local thickening of the front. Since welding of GFRP is not (yet) possible, the hull was bolted onto the aluminium alloy box beam vehicle frame. With **applique ceramic tiles** fitted, this hull was reported as having the same ballistic protection as a standard aluminium alloy M2 Bradley IFV but **weighing 27% less**. The density of GFRP is about 1,500 kg m^{-3} compared with 2,700 kg m^{-3} for that of aluminium alloy, making GFRP armour plates about twice the thickness of aluminium alloy plates for the same areal density. Reduced interior noise level and lower radar signature are two claimed advantages of GFRP over metal. However, on the debit side the augmenting ceramic tiles would shatter when hit (thus offering only 'one hit protection'), and the GFRP must be robust enough not to crack (particularly around the bolts to the alloy frame) especially when the vehicle is air-dropped to the ground. *Despite continuing research and development work by several agencies, a vehicle of this type is not yet in service.*

10 Alloys for Military Bridges

Since World War II the evolution of military bridges has been driven by the increasing requirement for rapid deployment. The need for transportability and ever quicker build-times has inevitably led to the use of higher strength to weight ratio alloys. Their weldability and fracture toughness are important considerations, the latter particularly so because of the possibility of battle damage - *surface notching from bullets and HE fragments is something the civil bridge designer does not have to worry about!*

The **Tables at the end of this chapter** enable comparison of the relevant materials properties and military bridge data.

MILD STEEL - BAILEY BRIDGE AND HEAVY GIRDER BRIDGE

These classic military bridges, seen in Plates 75 and 76, were fabricated in **mild steel** (0.25%C). Though having only a modest strength to weight ratio mild steel is inexpensive, easy to weld, and very 'forgiving' in use:

It has a **high tolerance to defects**. The critical defect size **cds** calculates to about 90 mm (worst case buried defect '2a') from **fracture toughness** theory-
The Griffith equation: $K_{Ic} = Y\sigma(\pi a)^{1/2}$ where:

K_{Ic} is fracture toughness
Y is a geometric (compliance) factor
σ is the working stress
a is the critical crack depth (cds) for
a surface defect

Taking the working stress as the yield stress of 350 MPa, its K_{Ic} value of 130 MPa m$^{1/2}$, and Y as 1, this gives the '**yield before break**' criterion at $2a = 90$ mm.
This is greater than the thickness of the girders, so that even a sharp edged full girder thickness crack (which since it is less than this critical size will give yielding before breaking) can be tolerated without fear of catastrophic brittle fracture. *Minor battle damage is not a problem here!*

It can also tolerate **extensive plastic buckling** before there is any danger of fracturing - as indicated by a high tensile ductility of around 35 %El.

And in a non-buckled structure **fatigue crack growth is unlikely** to have occurred - the fatigue stress value for 10,000 loading cycles (crossings) is not much below tensile yield stress, and so the presence of a crack would raise local stress to above the yield stress then giving visible plastic buckling.

ALUMINIUM ALLOY – MEDIUM GIRDER BRIDGE AND BR 90

In the mid-1960's the (then) Military Vehicle Experimental Establishment, MVEE Christchurch, developed aluminium alloy 'DGFVE 232' specifically to give a much lighter bridge structure with easier and speedier construction. This alloy was first used for the medium girder bridge **MBG** - seen in Plate 77 with a 65 tonne Chieftain main battle tank crossing. It was designed in man portable 1.8 metre long box sections (Plate 78) that could be joined together by pins to give a maximum span of 30.5 metres. A full span bridge can be assembled by 25 men in 90 minutes. For long spans and heavy loads the side girders may be deepened by adding an 'N' truss second storey – 'double storey' construction (Plate 79).

Alloy DGFVE 232 is Al-4Zn-2Mg with added Mn and Zr - a close relative of Scorpion armour alloy type 7039. The hot rolled or extruded plates are solution heat treated at 450°C and **quenched by forced air cooling**, then precipitation hardened in stages: 3 days at room temperature, followed by 8 hours at 90°C and then 16 hours at 150°C. This procedure ensures the lack of precipitate free zones PFZ close to the grain boundaries, significantly improving resistance to stress corrosion cracking but at the expense of strength. Typical tensile UTS is 390 MPa, compared with a value of 475 MPa for armour alloy type 7039.

Fatigue crack growth of the more highly stressed bridge parts has to be monitored carefully, since worst case buried critical defect size is about 9 mm compared with the mild steel value of 90 mm. **Stiffness** (Young's modulus E) of aluminium alloys is only one third that of steel on an absolute basis, but the bending stiffness of a structure is proportional to Et^3 where t is the thickness of the deflecting member. For matched fatigue strength designs aluminium alloy beams are thicker than those in steel thus nearly compensating for the low E of the material and yet still saving weight. Plate 80 shows a medium girder bridge with deflection limiting spars fitted on the tensile underside to help increase bending stiffness.

The medium girder bridge replacement called 'BR 90' (Plates 81 and 82) is also made in DGFVE 232 aluminium alloy. Deployment of the sections is more highly mechanised using launch rails, and this is called 'mechanically aided construction by hand' or **MACH**. The long span 55 metre variant uses the lightest possible launch rails made in aluminium alloy type 7075 and/or carbon fibre reinforced polymeric CFRP.

MARAGING STEEL – ARMOURED VEHICLE LANCHED BRIDGE

In the late 1960's Christchurch designed the 'battle group' armoured vehicle launched bridge **AVLB**, the 24.4 metre long bridge being carried folded on the Chieftain tank hull in place of the gun turret. The ingenious "praying mantis" launching sequence (taking only 5 minutes) is sketched below:

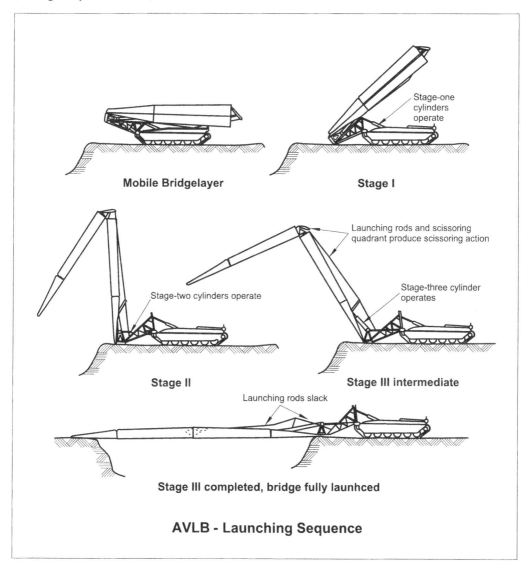

Mobile Bridgelayer

Stage-one cylinders operate

Stage I

Launching rods and scissoring quadrant produce scissoring action

Stage-two cylinders operate

Stage-three cylinder operates

Stage II

Stage III intermediate

Launching rods slack

Stage III completed, bridge fully launhced

AVLB - Launching Sequence

Plate 83 shows the bridge being deployed. Plate 84 shows the bridgelayer vehicle crossing its own bridge after uncoupling. This it does when the main battlegroup has crossed, before then turning round to pick the bridge up again and carrying on to the next gap along the road.

The original armoured vehicle launched bridge design was in aluminium alloy DGFVE 232, but at 21 tonnes unacceptably heavy. Ultra-high strength **maraging steel** was then selected to achieve a weight of about 12 tonnes, much nearer the weight of the turret it replaces on the vehicle - it would now not slow the battlegroup down. *This must be one of the largest structures made in this expensive steel, which is more usually found in aircraft undercarriage components for example.*

The word **maraging** is short for 'martensitic age hardening'. It is a high alloy steel 18Ni8Co, but with a very low carbon content 0.03%C, and also contains Mo, Ti and Cu to give age hardening precipitates.

Heat treatment is:

(1) Solution anneal for 30 minutes at 820°C then air cool. The high alloy content ensures air hardening to martensite, but because of the low carbon content it is only about 300 Hv hardness - so called 'ductile martensite'.

(2) Precipitation harden for 3 hours at 480°C. This strengthens the steel to the 1500 MPa tensile yield stress level.

Fabrication (bending, cutting, drilling) is carried out in the solution annealed condition, impossible after the fourfold strength increase resulting from precipitation hardening - an excellent solution to the problem of how to form to shape an ultra-high strength component.

After any subsequent **repair MIG welding** (which would re-solutionise the all important precipitates) the weld heat affected zone is re-aged at 480°C using local electrical 'blanket' heaters.

Maraging steel was developed with very high **fracture toughness** in mind - by double vacuum re-melting to minimise the incidence of non-metallic inclusions, also giving a **high fatigue strength**. But even so, worst case buried critical defect size is rather small at 5 mm, since yield stress is very high. However, in the 1980's main battle tank engine power significantly increased and the planned **AVLB replacement** can now be heavier. This allows the decking to be made in aluminium alloy, thus reducing costs and at the same time alleviating fears regarding battle damage critical defect size.

POSSIBLE ALTERNATIVE ALLOYS AND CFRP - FUTURE BRIDGES

Other high strength to weight ratio alloys such as aircraft aluminium alloy type 7075 and titanium alloy Ti-6Al-4V (IMI 318) might be considered for future bridges beyond BR 90. *Their mechanical properties are detailed in the Tables at the end of this chapter*: The use of **aluminium alloy type 7075** would not save much weight compared to DGFVE 232, since despite a 45% higher strength/weight ratio its fatigue strength is much the same. Also its higher yield strength means that its worst case critical buried defect size would be a very worrying 2 mm.

The use of **titanium alloy Ti–6Al–4V** would give improved stiffness and save around 25% structural weight compared with aluminium alloys. And compared with maraging steel its fatigue strength is at least as good, together with a similar critical defect size (5 mm). So that titanium alloy does show potential for this application.

For future bridges of the BR 90 type, the graph right (*DERA Chertsey*) shows the possible weight advantages and relative costs of these two alloys.

It also shows that an all **CFRP** bridge might be half the weight of DGFVE 232 aluminium alloy, but at twice the cost. Longitudinally reinforced CFRP (uniply) has a similar stiffness and UTS to maraging steel at about 20% of the weight, giving exceptional strength/weight ratio.

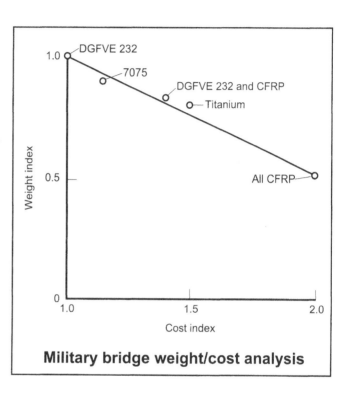

Military bridge weight/cost analysis

The possible use of **CFRP** for military bridges has been researched for several years and is not a simple problem. Limit of tensile linearity (yield stress in a metal) is around 900 MPa, although fibre pull-out or resin cracking can occur at a somewhat lower stress. A threshold for non-damage is sometimes taken at about half this value giving an effective 'fatigue stress' of around 400 MPa. These **properties are directional** being substantially lower in transverse 'across the fibres' tests. Strain to fail is only 1.5% (El) so that plastic buckling is non-existent, and work hardening will not occur either. Fracture toughness is around 40 MPa m$^{1/2}$ longitudinally, (similar to DGFVE 232) and if CFRP were a metal then the Griffith equation would give a critical defect size of only 1 mm. So barely visible impact damage **BVID** is of prime concern - not boding well for military robustness. These negative factors plus the need for jointing with relatively low strength **adhesives** present a considerable challenge to the bridge designer. If deflections are designed to be much less than for metals (by sacrificing some of the considerable weight advantage to thicken members, for example) then the scope for 1 mm internal defects forming is reduced, though **battle damage** to the tensile underbelly is a worrying constraint.

However, as user experience of CFRP builds into a better understanding of its failure modes, and as section thicknesses routinely increase from current racing car and aircraft panel gauges then these worries will subside. The potential is for a BR 90 type bridge at around 6 tonnes in weight with no other currently available material coming close to that. *This in turn opens up possibilities for rapidly deployed longer span bridges.*

SOME MILITARY BRIDGE DETAILS

Bridge	weight (tonnes)	span (m)	mlc (tonnes)	construction	comments	Material
BB	80	30	60	90 men 8hrs	all steel	MILD STEEL (0.25%C)
HGB	90	30	60	24 men +crane, 5hrs	Al alloy deck and 2 way traffic	
MGB	21.3	30	60	25 men 1.5 hrs, or MACH	double storey usual	ALUMINIUM ALLOY
BR 90	12	32	70	10 men 0.5hrs	all Al alloy	Al-4Zn-2Mg
	25	55	70	MACH with launch rails	launch rails 7075 or CFRP	
AVLB first design	21	30	60	-		DGFVE 232
AVLB in service	12.2	24.4	60	vehicle 5 mins		MARAGING STEEL
AVLB replacement	15?	24.4	60	vehicle 5 mins	Al alloy deck	18Ni-8Co
Beyond BR 90 ?						ALUMINIUM ALLOY 7075 Al-6Zn-2Mg
						TITANIUM ALLOY Ti-6Al-4V
	6? half weight, 2X cost of Aluminium	32	70	MACH with launch rails	longer spans?	CFRP
						ADHESIVE (Hysol 9309)
	mlc is Max Load Capacity		MACH is Mechanically Aided Construction by Hand			

SOME TYPICAL PROPERTIES OF BRIDGE MATERIALS

Material	E	ρ	Tensile			Strength to weight ratio	Fatigue stress for 10^4 cycles	Fracture toughness K_{Ic}	Cost index per tonne
			YS	UTS	El				
	(GPa)	(sg)	(MPa)	(MPa)	(%)	UTS/ρ	(MPa)	(MPa m$^{1/2}$)	
MILD STEEL (0.25%C)	210	7.9	350	500	35	54	300	130	0.1
Al ALLOY DGFVE 232	70	2.8	340	390	15	139	210	40	1
MARAGING STEEL	180	8.4	1400	1460	18	174	500	110	4
Al ALLOY 7075	71	2.8	500	570	12	204	200	30	1.2
Ti ALLOY Ti-6Al-4V	110	4.4	900	950	15	216	580	75	6
CFRP (uniply)	200 (dir)	1.5	900 (dir)	1400 (dir)	1.5	930	400? (dir)	40?	8 (dir)
Hysol 9309	0.7 (shear)	-	-	30 (shear)	-	-	-	-	-

raw matl.

(dir) is directional

11   Alloys for Gun Carriages and Tank Track Links

105 mm LIGHT GUN TRAIL

The 105 mm light gun is photographed in Plates 85 and 86, and sketched right.

The high **recoil** force of any large gun puts considerable stress on the trail legs during firing, and their **fatigue** life is important, since several thousand firing cycles are expected without failure.

This consideration combined with the requirement for **air-portability** led to the selection of '**FV520**' **ultra-high strength high alloy steel** (sometimes known as 'STA 59' in defence circles) for these trail legs.

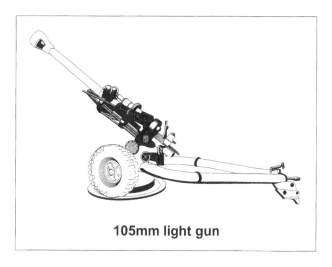

105mm light gun

Four trail sections are each made by progressively cold drawing down a centrifugally cast tube. Drawing is commenced in the austenitic condition, and the rapid work hardening rate of this material provides the high tensile strength necessary immediately below the die to prevent tearing during drawing. Annealing at 1050°C is necessary after each draw. Each of the two trail legs is then fabricated by **electron beam welding** two drawn sections together - done in a vacuum tank with no filler, giving a high integrity 'clean' weld. The seam can be seen in the sketch above about halfway along each leg. Finally heat treatment is carried out as detailed below.

FV520 is a semi-austenitic controlled transformation steel. It is a high alloy steel 16Cr6Ni, with a low carbon content 0.05%C, and also contains Mo, Ti and Cu to give age hardening precipitates.

Heat treatment is:

(1) Solution anneal for 30 minutes at 1050°C then air cool. The Cr and Ni contents are accurately balanced (in each cast of steel) so that on air cooling the microstructure

is mainly austenite - with less than 10% interspersed martensite grains. The martensite start temperature M_s is around 50°C. *Fabrication is carried out in this condition.*

(2) 'Condition' for 2 hours at 750°C then air cool. This allows carbide precipitates to form, lowering the %C in the grains and raising their M_s temperature. Then during air cooling the microstructure is about 90% martensite and 10% austenite - since the martensite finish temperature M_f is still below 0°C.

(3) Refrigerate for 2 hours, usually at below minus 25°C. This takes the component down to below the M_f temperature, transforming any retained austenite to martensite.

(4) Precipitation harden for 2 hours at 450°C. This strengthens the steel to around 1100 MPa tensile yield stress - not quite as strong as maraging steel, but with excellent fatigue strength (540 MPa for 10^5 cycles) and at about half the cost.

This is a complex heat treatment schedule, and some would say "a quality controller's nightmare" - not only is the chemical analysis finely balanced, but the mechanical properties after each stage have to be correct ready for the next stage.

155 mm FH 70 Gun Trail

The trail legs of the much larger 155 mm FH 70 gun (Plate 87) are also fabricated in **FV520 steel,** but instead of being tubular they are of welded box section construction.

155 mm UFH Gun Trail

The 155mm ultra-light weight field howitzer (Plate 88) can be air-lifted by a single main rotor helicopter, such as the Black Hawk, instead of the double main rotored Chinook required to lift FH 70. The lighter weight also means there is less chance of 'bogging down' in wet sand during a beach landing. The weight is saved by constructing the trail and some other carriage parts in **titanium alloy Ti-6Al-4V** (IMI 318). The tensile yield stress of this alloy is about 900 MPa compared with 1080 MPa for FV520 steel, but its density is only 4,400 kg m^{-3} (compared with 8,400 kg m^{-3} for FV520 steel) so that its yield strength to weight ratio or **specific yield strength** is 60% higher. In addition titanium alloy Ti-6Al-4V has a similar absolute fatigue stress to FV520 steel - around 580 MPa for 10^4 stress cycles - giving it a 48% higher **specific fatigue strength**. However, titanium alloy Ti-6Al-4V is about three times the cost of FV520 steel per ingot tonne.

MAIN BATTLE TANK TRACK LINKS AND PINS

A main battle tank track link, such as that fitted to Chieftain sketched right, is sand cast in **13%Mn Hadfield steel** - with 1%C, and at first sight is perhaps a surprising choice.

This (unusual) steel is also used for bulldozer blades, rock carrying buckets, and other excavator components. The common denominator is the requirement for hard, **wear resistant** bearing surfaces - such as the 'horn' on the track link which rubs between the double road wheels, and the flat lower surfaces in contact with the

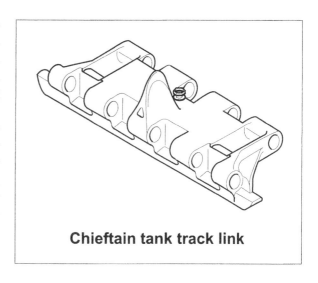

Chieftain tank track link

road. After casting, the link is heat treated for 1 hour at 1030°C then water quenched. Manganese (like nickel) is an **austenite** stabiliser, and the resulting microstructure is FCC austenite grains.

During plastic deformation, FCC austenitic steels (the most common example being 18Cr-8Ni stainless steel) give **a higher rate of work hardening** than BCC ferritic or martensitic steels - as shown in this tensile curve. Hadfield steel gives a particularly pronounced effect, with hardness rising from **180 Hv to about 550 Hv** during work hardening. During early use initial 'running in', the track horn and lower surface will work harden appropriately - an active 'smart' response to their service conditions. *The old 'tin helmets' worn by soldiers in battle were also made in this steel, giving excellent bullet resistance by the same mechanism.*

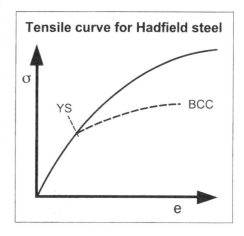

Tensile curve for Hadfield steel

σ

YS

BCC

e

The **track pins** are made in **low alloy steel** - 1´%NiCrMo BS970 - 817M40 grade (En24), oil quenched and fully tempered to the UTS 1150 MPa level. They are then **induction hardened** - this involves rapid surface heating of the pins in an induction coil, then water quenching to give a martensite (hard) 'case' around 150 microns deep. This is to give the highly loaded pins decent wear resistance in the track link holes. Even so they can sometimes fail after only a few tens of miles, and spare track link 'wraps' (each containing 8 links and pin sets) are commonly carried on main battle tanks often bolted to the outside rear of the hull.

12 Dynamic Behaviour of Alloys at High Strain Rate

Ammunition components and armours are obviously expected to function at high rates of strain. For instance the long rod kinetic energy penetrator strikes the target at around 1500 m s^{-1} and a shaped charge jet can impact at speeds as high as 10 km s^{-1}, equating to strain rates $\acute{e}$ in the region of 3.10^3 s^{-1} and 10^5 s^{-1} respectively. The conventional tensile test using a **tensometer** (Plate 2), as detailed in chapter 1, returns a strain rate of around 10^{-3} s^{-1}. This is a quasi-static or **static** test some six orders of magnitude slower than ballistic events. More realistic **dynamic** tensile testing can be performed on an ultra-fast servohydraulic tensometer, or on an **instrumented drop tower** (Plate 89) fitted with a tensile attachment (Plate 90) giving much higher crosshead speeds .

EFFECT OF STRAIN RATE ON MECHANICAL PROPERTIES

Generally **as strain rate is increased metals get stronger but less ductile** - as seen right.

This effect is similar to lowering the temperature by using an environmental chamber around the tensile specimen. For steel increasing the strain rate by a factor of 10^3 s^{-1} is equivalent to lowering the temperature by about 50°C.

Interestingly, the elastic stiffness (Young's modulus E) is insensitive to strain rate for metals, whereas it can vary quite considerably with strain rate for many non-metals. Also the area under the tensile curve (the energy to fracture) may not change much with strain rate.

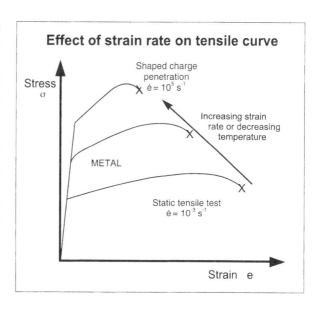

The ammunition or armour designer can rely on higher strength values than those obtained from static tensile testing, but there is less capacity for plastic deformation at higher strain rate and the problem is to avoid premature fracture because of reduced ductility.

The Ludwig equation (sometimes called the Holloman equation) relates true stress σ to true strain ε and strain rate $\varepsilon^{\cdot}$ during plastic deformation:

$$\sigma = \sigma_0 + K\varepsilon^n\varepsilon^{\cdot m} \quad \text{where } \sigma_0 \text{ and } K \text{ are alloy constants,}$$
$$n \text{ is the work hardening index and}$$
$$m \text{ is the strain rate sensitivity index (both alloy constants).}$$

At constant plastic strain, ε^n is constant and can be incorporated as constant K_1. If that plastic strain is 0.2%, then the corresponding stress will be the 0.2%PS which is nearly the same as YS (σ_y), and the equation reduces to:

$$\sigma_y = \sigma_0 + K_1\varepsilon^{\cdot m}$$

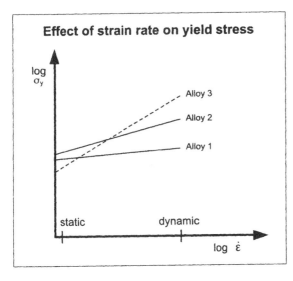

So a plot of $\log \sigma_y$ against $\log \varepsilon^{\cdot}$ will be a straight line with slope m, as shown left for three different alloys.

The strain rate sensitivity index m varies with different alloys and also with different microstructures of the same alloy.

A low static yield strength alloy (alloy 3) will have a high dynamic yield strength, if it has a high strain rate sensitivity index.

Strain rate sensitivity often relates to crystal structure - HCP metals (Zn or Mg for instance) generally give higher m values than BCC metals (Fe say), and FCC metals such as Cu and Al usually give the lowest values.

The 'league table' of materials strengths embedded in many engineers minds is the static data, and the 'dynamic league table' is quite different.

This fact is now being increasingly realised, and is important not only to the military designer. **Even in the civil sector** the static data used by most design engineers for materials selection is often inappropriate for the application in mind. The strain rate at the the tool tip during machining is very high; the railway line deforms fast under the train wheels; in vehicle crashworthiness testing the structure crumples at high speed; engine and drivetrain components work at high strain rate; the list could go on. Also in the increasingly important areas of **mathematical modelling** and computer aided design, the material coefficients should often be dynamic rather than static. The computer model will iterate the material flow ad infinitum - into the realms of pure fantasy - unless a boundary condition (strain to fail) is set, and this should clearly be the value at the strain rate appropriate to the event being modelled.

*As well as investigating dynamic tensile testing, researchers are also looking at **dynamic compression** testing and **dynamic fracture toughness** testing - K_{Id} instead of K_{Ic} . The European Structural Integrity Society ESIS, for example, is currently coordinating work towards European test method standards for all three modes.*

DEFORMATION TWINNING IN STEELS

Plastic deformation at conventional (low) strain rates, say by cold rolling, gives elongated work hardened grains. Above the yield stress lattice **dislocations** move, multiply, and then get in the way of each other impeding further movement (the mechanism of work hardening). At high strain rates the dislocations have less time to move, and so ductility is curtailed giving premature fracture.

Explosive shock loading gives elongated grains too, but in BCC steels we often also see **deformation twins** - sometimes called Neumann bands. These can be seen in pure iron in Plate 91, and in the ferrite grains of shock loaded mild steel in Plate 92. They get thinner and more numerous at higher strain rates as shock hardening increases. It is difficult for dislocations to move faster than 2000 m s⁻¹ and this second deformation mechanism is then invoked. Deformation twins can sometimes be observed in BCC steels deformed at slow strain rates, but only if the temperature is very low (say at minus 196°C, liquid nitrogen temperature).

The mechanism of twinning is sketched right. The twin boundaries (dotted lines) act as mirror planes of the lattice orientation - which is normally constant within a single grain.

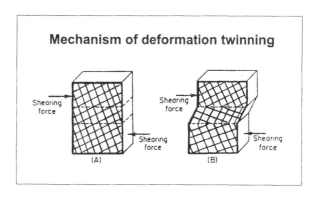

Deformation twinning has not been observed in the more ductile FCC metals, but twinning can occur in these during annealing heat treatment, giving **annealing twins** - as seen in cartridge brass (Plate 19) for instance.

ADIABATIC HEATING EFFECTS

At high strain rate there is little time for the heat generated by plastic deformation (mechanical work or Joule heating) to dissipate, and the test specimen or component gets hot - 'adiabatic heating'. This is nicely illustrated when performing **compression tests** on small solid cylinders of tungsten penetrator alloy between plane platens -

static testing on an Instron machine with initial strain rate of 10^{-3} s^{-1} and dynamic testing on an instrumented drop tower with initial strain rate of about 10^3 s^{-1} :

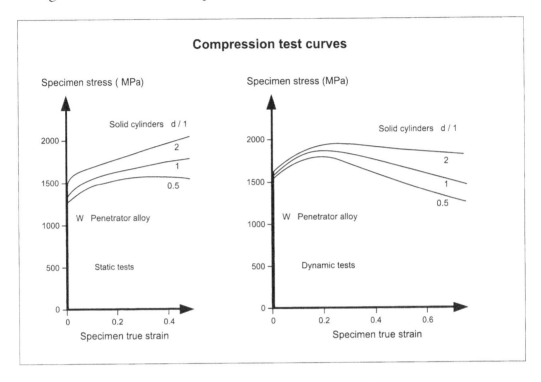

The static curves (left) show conventional work hardening, but in the dynamic tests (right) **thermal softening** due to adiabatic heating causes the curves to drop off.

Testing cylinders with different diameter to length ratios (d/l) enables the cylinder end-face friction forces to be quantified and then subtracted - the Cook and Larke method. The shorter fatter cylinders give artificially higher stress values because of greater end-face friction.

All metals show lower strength at higher temperatures, but thermal softening can also be concentrated locally in zones of intense shear stress for instance. This can give rise to **adiabatic shear bands** of changed microstructure as shown in the micrographs of Plates 93 to 96 - in medium carbon steel, aluminium alloy, titanium alloy, and depleted uranium alloy respectively. Once shearing starts it concentrates the plastic deformation in the bands and runaway failure can occur - in a dynamic compression test this can be seen as a sudden drop in the stress/strain curve. This is one way in which **plugging failure** can occur in a target penetrated by a blunt or soft penetrator (Plates 93 and 95) as detailed in chapter 5.

The critical shear strain γ for the formation of adiabatic shear band is given by the **Culver equation**:

$$\gamma = \frac{C\rho n}{\delta\tau/\delta T}$$

where C is heat capacity per unit mass
               ρ is density
               n is work hardening index
               $\delta\tau/\delta T$ is rate of change of shear yield
                    strength with temperature

For adiabatic shear bands not to occur γ should be as high as possible - with terms $C\rho n$ being as high as possible, and thermal softening $\delta\tau/\delta T$ being as low as possible.
Examination of these properties shows titanium alloys and depleted uranium alloys to be particularly susceptible. Depleted uranium alloys have a high thermal softening rate combined with a phase transformation at about 600°C.

The whole field of dynamic properties and behaviour of alloys at high strain rate is an important and fascinating one, and there is much to learn yet. There is even more to learn about these aspects in non-metallic and composite materials.

TYPICAL MECHANICAL PROPERTIES FOR A SELECTION OF ALLOYS AND MATERIALS

ALLOY OR MATERIAL	E Youngs Modulus (GPa)	ρ Density (kg/m³)	Hv Hardness (Hv)	YS Yield Stress (MPa)	UTS Max Stress (MPa)	S Fatigue Stress 10^7 cycles (MPa)	%El Elongation to Fracture (%)	Impact Energy to Fracture at +20°C (Charpy J)	K_{Ic} Fracture Toughness (MPa m$^{\frac{1}{2}}$)	T_m Melting Temp (°C)	Cost index per tonne (ratio)
Aluminium 1100 Annealed	69	2710	24	40	90	34	45	-	-	643	8
Al-Mg 5083 Annealed	70	2660	68	145	290	-	22	-	-	574	8
Al-Mg 5083 H34	70	2660	105	285	345	160	9	-	-	574	8
Al-Cu 2014 T6	72	2800	140	414	483	125	13	-	31	507	10
Al-Zn-Mg 7039	70	2780	140	400	470	-	15	20	-	535	12
Al-Zn-Mg-Cu 7075 T6	71	2800	157	503	572	159	11	-	28	532	12
Al-12Si As Cast	71	2660	68	145	295	130	3	-	-	575	6
Mg alloy AZ91B As Cast	45	1810	66	150	230	97	3	3	20	470	11
Cu OFHC Annealed	118	8940	45	62	216	-	60	43	-	1083	7
Cu-30Zn 70/30 Brass Annealed	105	8520	65	78	325	114	65	58	-	930	7
Cu-30Zn 70/30 Brass CW 50%	105	8520	200	386	575	150	2	-	-	930	7
Cu-40Zn 60/40 Brass Extruded	105	8390	150	385	540	-	10	-	-	900	7
Ti-6Al-4V Annealed	110	4430	370	825	895	500	10	27	75	1604	60
Mild Steel AC 040A20-En2 0.2%C	210	7850	127	280	430	220	40	100	135	1490	1
1½%Mn Steel WQ+T650°C 0.2%C 150M19-En14A	210	7850	216	570	700	430	27	100	186	1490	1.1
1%CrMo Steel WQ+T650°C 0.4%C 708M40-En19A	215	7830	272	770	880	460	25	145	-	1500	1.4
1½%NiCrMo Steel WQ+T650°C 0.4%C 817M40-En24	210	7840	288	790	930	490	24	105	-	1510	1.6
3%NiCrMo Steel WQ+ T650°C 0.3%C 830M31-En27	207	7860	314	970	1050	510	19	122	150	1530	2.2
BM1 Tool Steel [8Mo4CrCo] 0.8%C	211	7750	850	1700	2100	-	0.5	10	-	1500	5
19/10 Stainless Steel 19Cr10Ni austenitic AISI 304 0.06%C	190	8000	190	200	600	-	35	80	-	1600	4
Maraging 1400 Steel fully aged 0.03%C 18Ni8Co5MoTi	180	8370	450	1400	1460	450	14	27	145	1430	40
Maraging 2400 Steel fully aged 0.03%C 18Ni12Co4Mo2Ti	194	8390	650	2400	2430	450	9	6	45	1440	55
Polypropylene	1	900	7	32	33	-	600	-	-	150	2.7
Nylon 6-6	2	1240	7	57	80	-	200	-	-	200	7.5
Polycarbonate	2	1200	10	-	60	-	125	-	-	165	4.5
Silicon Carbide	350	3200	2700	250	250	-	-	-	4	1600	8
Silicon Nitride (hot pressed)	160	3100	2000	400	400	-	-	-	5	1650	40
Boron Carbide	460	2500	2800	180	180	-	-	-	-	1200	40

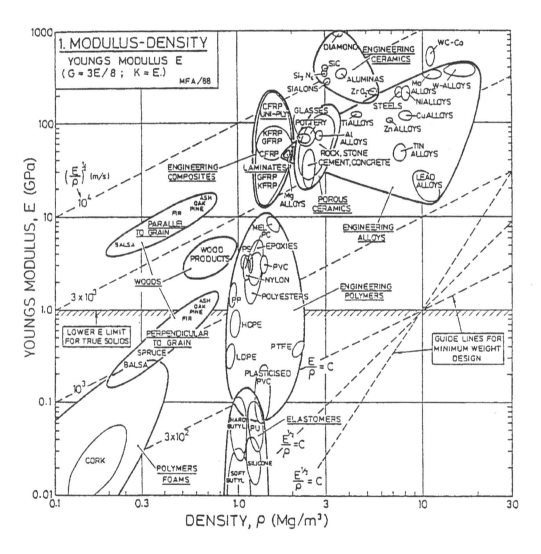

Ashby Materials Selection Diagram   -   MODULUS - DENSITY

M F Ashby: Acta Metallurgica 1989 37 1273

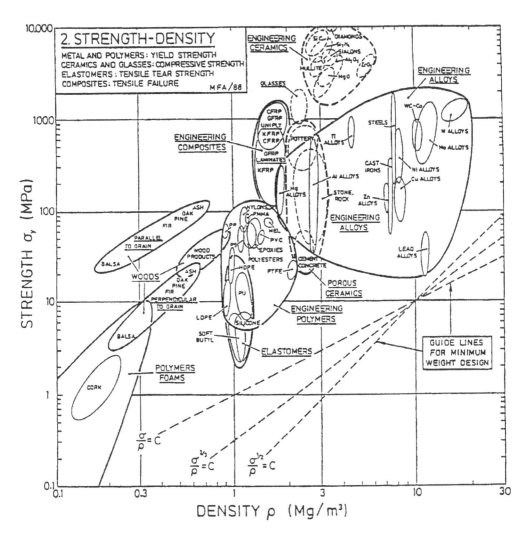

Ashby Materials Selection Diagram -   STRENGTH - DENSITY
*M F Ashby: Acta Metallurgica 1989 **37** 1273*

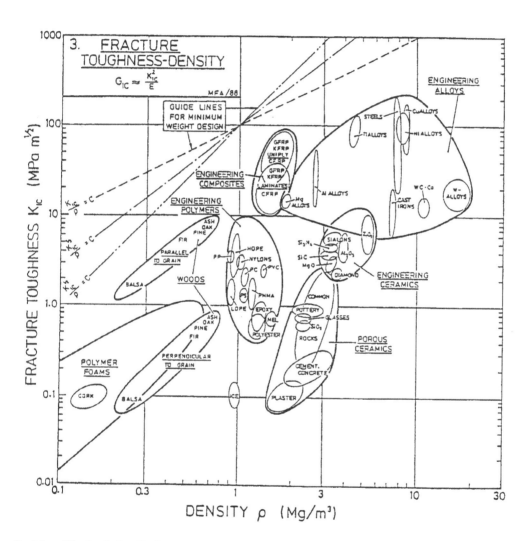

Ashby Materials Selection Diagram   -

FRACTURE TOUGHNESS - DENSITY

*M F Ashby: Acta Metallurgica 1989 **37** 1273*

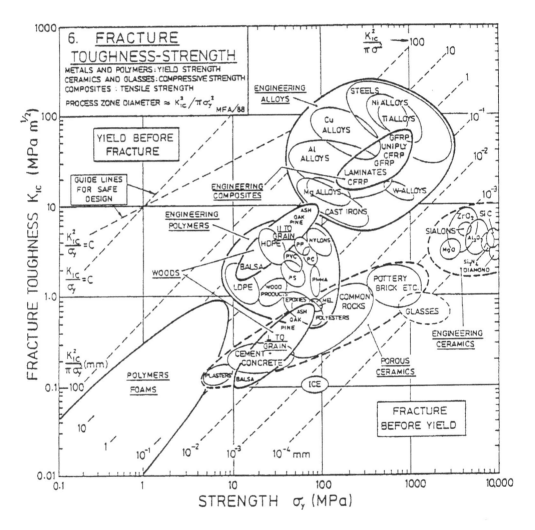

Ashby Materials Selection Diagram   -   **FRACTURE TOUGHNESS -**
STRENGTH

M F Ashby: Acta Metallurgica 1989 37 1273

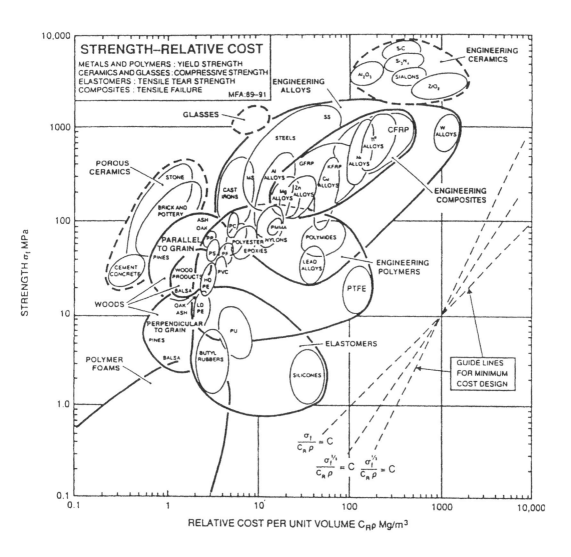

Ashby Materials Selection Diagram -   **STRENGTH -**
                                            RELATIVE COST

M F Ashby

CHEMICAL ELEMENTS

[an alphabetical order list of symbols used in this book]

symbol	element	atomic number	atomic weight	symbol	element	atomic number	atomic weight
Ag	silver	47	108	Mn	manganese	25	55
Al	aluminium	13	27	Mo	molybdenum	42	96
Ar	argon	18	40	N	nitrogen	7	14
Au	gold	79	197	Nb	niobium	41	93
B	boron	5	11	Ni	nickel	28	59
Be	beryllium	4	9	O	oxygen	8	16
Bi	bismuth	83	209	Pb	lead	82	207
C	carbon	6	12	Pt	platinum	78	195
Ca	calcium	20	40	S	sulphur	16	32
Ce	cerium	58	140	Sb	antimony	51	122
Co	cobalt	27	59	Si	silicon	14	28
Cr	chromium	24	52	Sn	tin	50	119
Cu	copper	29	64	Ta	tantalum	73	181
DU	depleted uranium	92	238	Ti	titanium	22	48
				U	uranium	92	238
H	hydrogen	1	1	V	vanadium	23	51
Fe	iron	26	56	W	tungsten	74	184
La	lanthanum	57	139	Y	yttrium	39	89
Li	lithium	3	7	Zn	zinc	30	65
Mg	magnesium	12	24	Zr	zirconium	40	91

ALLOY COMPOSITIONS IN THIS BOOK

For alloy chemical compositions **% is by weight**, unless otherwise stated. **Non-ferrous alloys** are often shown like **Ti-6Al-4V** for example, a titanium alloy containing 6% by weight of aluminium and 4% by weight of vanadium.

STEELS SHORTHAND NOTATION IN THIS BOOK

Plain carbon steels are shown like **0.2%C mild steel** for example, indicating the weight % carbon, the balance of the composition being mainly iron. Most **low alloy steels** are expressed like **3%CrMoV steel** for example, where the principal alloying element is chromium at 3% by weight. The other two alloying elements are present at less than 1% by weight, but there is more molybdenum than vanadium in the steel. The balance of the composition is carbon (usually less than 1% by weight), some residual elements (such as sulphur and silicon), but mainly iron. **High alloy steels** are usually written in the same way as the non-ferrous alloys, see above.

SOME FURTHER READING

Military Technology

Jane's Defence Guidebooks, Jane's, Coulsdon, Surrey:

C. Foss: *Armour and Artillery*, 17th Edition 1994/5,
ISBN 0-7106-1374-1.
C. Foss and T. Gander: *Military Vehicles and Logistics,* 17th Edition 1995/6,
ISBN 0-7106-13504.
T. Gander and I Hogg: *Ammunition Handbook*, 5th Edition 1996/7,
ISBN 0-7106-13784.

Brassey's Battlefield Weapons Systems and Techology Series, London:

I. Tytler et al: *Vehicles and Bridging*, Series 1 Vol.I 1985,
ISBN 0-08-028325-3.
M. Manson: *Guns, Mortars and Rockets*, Series 3 Vol.3 1997,
ISBN 1-85753-172-8.
P. Courtney-Green: *Ammunition for the Land Battle*, Series 2 Vol.4 1991,
ISBN 0-08-035807-1.
T. Terry et al: *Fighting Vehicles*, Series 2 Vol.7 1991,
ISBN 0-08-036704-6.
D.Allsop: *Cannons*, Series 3 Vol.2 1995,
ISBN 1-85753-104-3.

I. Hogg: *The Illustrated Encyclopedia of Ammunition*, Quarto Publishing, London, 1985,
ISBN 1-85076-0438.
I. Hogg: *The Illustrated Encyclopedia of Artillery*, Quarto Publishing, London, 1987,
ISBN 1-55521-310-3.
C. Chant: *Compendium of Armaments and Military Hardware*,
Routledge & Keegan Paul Ltd, London, 1987, ISBN 0-7102-0720-4.
R. Lee: *Defence Terminology*, Brassey's, London, 1991,
ISBN 0-08-041334-X.
W. Walters and J. Zukas: *Fundamentals of Shaped Charges*,
Wiley Interscience, Chichester, Sussex, 1989, ISBN 0-471-62172-2.

Regular Magazines:-

Jane's International Defence Review: Jane's, Coulsdon, Surrey,
ISSN 0020-6512.
Defence Systems International: Stirling Publications Ltd, London,
ISSN 0951-9688.
Military Technology: Wher and Wissen Ltd, Bonn,
ISSN 0722-326.

Metallurgy and Materials Science

D. Llewellyn: *Steels - Metallurgy and Applications*, 2nd edition 1994 (or later),
Butterworth Heinemann Ltd, Oxford,
ISBN 0-7506-2086-2.
R. Honeycombe and H. Bhadeshia: *Steels - Microstructure and Properties*, 2nd edition
1995, Edward Arnold Ltd, London,
ISBN 0-7131-2793-7.
G. Krauss: *Principles of Heat Treatment of Steel*, 5th edition 1988 (or later), American
Society for Metals, Ohio,
ISBN 0-87170-100-6.
I. Polmear: *Light Alloys - Metallurgy of the Light Metals*, 1st edition 1981
(or later), Edward Arnold Ltd, London,
ISBN 0-7131-2819-4.
R. Higgins: *Engineering Metallurgy*, 5th edition 1983 (or later),
Hodder & Stoughton Ltd, London,
ISBN 0-340-28524-9.
J. Lancaster: *Metallurgy of Welding*, 3rd edition 1980 (or later),
George Allen & Unwin Ltd, London,
ISBN 0-04-669009-3.

D. Askeland: *The Science and Engineering of Materials*, 3rd edition 1996
(or later), Chapman & Hall Ltd, London,
ISBN 0-412-53910-1.
F. Crane and J. Charles: *Selection and Use of Engineering Materials*,
3rd edition 1989 (or later), Butterworths Ltd, London,
ISBN 0-408-10859-2.
J. Martin: *Materials for Engineering*, 1st edition 1996, Institute of Materials, London,
ISBN 1-86125-012-6.

PLATES SECTION

Plate 1 - Tensile test specimens and Charpy impact test specimen.

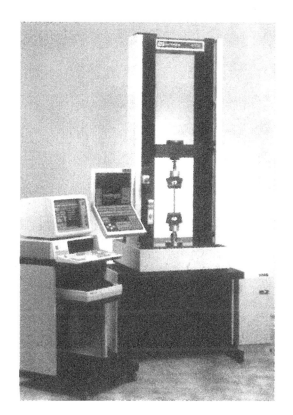

Plate 2 -
Tensile test machine. *Instron*

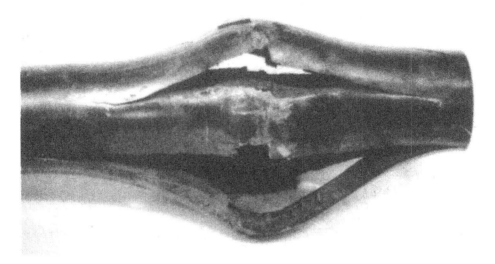

Plate 3 - General purpose machine gun barrel GPMG - ductile fracture.

Plate 4 - SS Schenectady - brittle fracture on a macro scale.

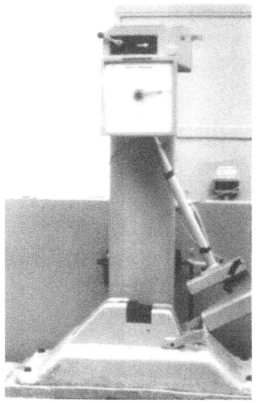

Plate 5 -
Charpy impact pendulum
machine. *Avery*

Plate 6 -
Vickers hardness test
machine. *Vickers*

Plate 7 -
Rockwell hardness test
machine. *Avery*

100μm

Plate 8 - Vickers hardness impression on cartridge brass.

Plate 9 - Optical microscope *Reichart-Jung*; Computerised image analyser.

Plate 10 -
Scanning Electron
Microscope SEM. *JEOL*

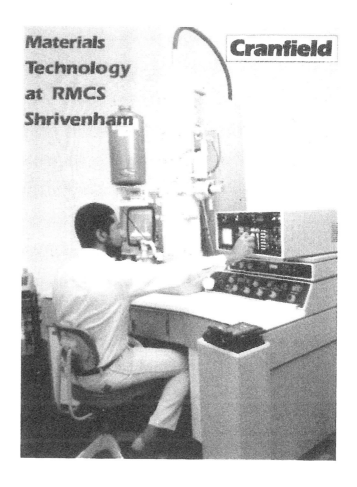

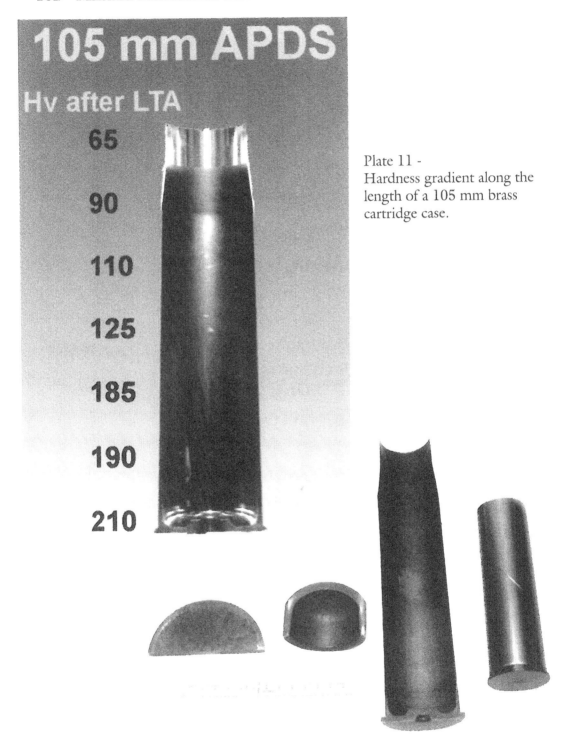

Plate 11 -
Hardness gradient along the length of a 105 mm brass cartridge case.

Plate 12 - 105 mm brass disc, cup and finished case; Wrapped steel case.

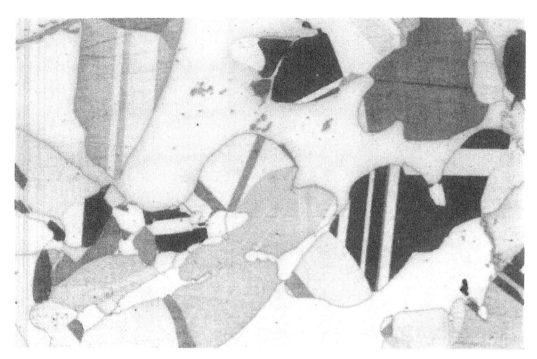

Plate 13 - 60/40 brass microstructure.

50μm

80μm

Plate 14 - 70/30 brass microstructure - annealed at 650°C for 30 minutes.

Plate 15 - 70/30 brass microstructure - cold rolled 50% [CR].

80µm

Plate 16 - 70/30 brass microstructure - cold rolled 50% [CR] at higher
magnification.

40µm

Plate 17 - 70/30 brass microstructure - CR then annealed at
      350°C for 30 minutes.

80μm

80μm

Plate 18 - 70/30 brass microstructure - CR then annealed at
      500°C for 30 minutes.

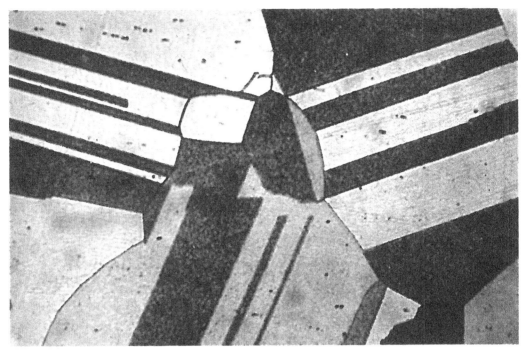

Plate 19 - 70/30 brass microstructure - CR then annealed at 750°C for 30 minutes.

80μm

Plate 20 - Stress corrosion cracking SCC in 70/30 brass.

60μm

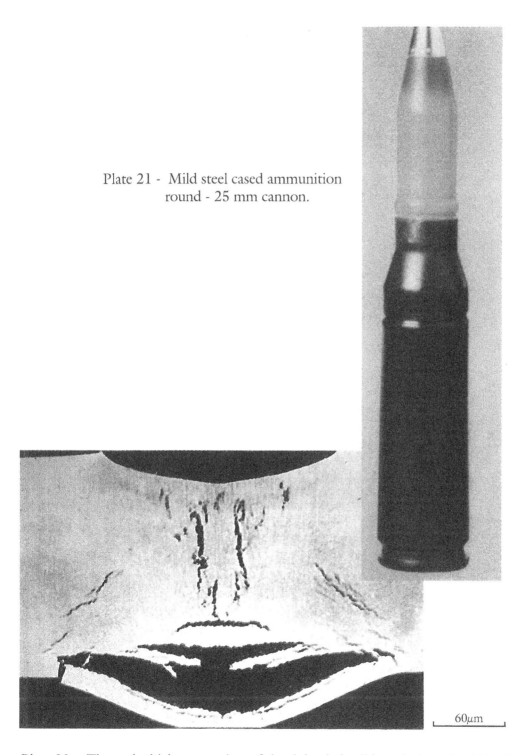

Plate 21 - Mild steel cased ammunition round - 25 mm cannon.

60μm

Plate 22 - Through-thickness section of shock loaded mild steel plate - 'scabbing'.

Plate 23 - 76 mm and 105 mm
        HESH steel projectile
        bodies.

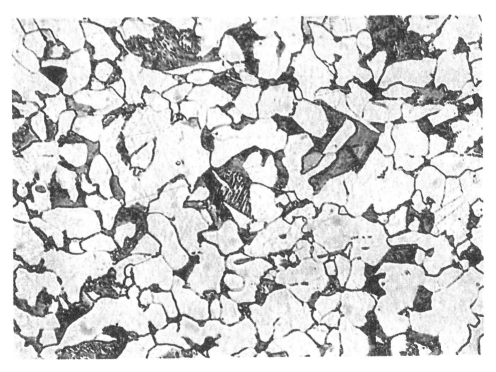

Plate 24 - 0.2%C steel microstructure - air cooled from 860°C

60μm

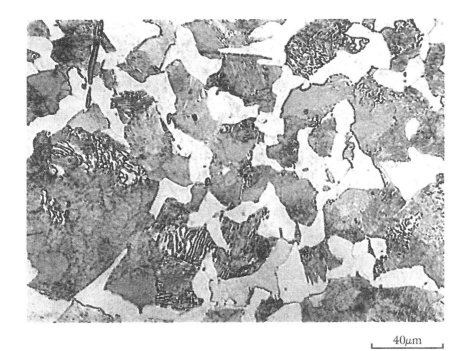

40µm

Plate 25 - 0.4%C steel microstructure - air cooled from 860°C

30µm

Plate 26 - 0.8%C steel microstructure - water quenched from 860°C.

Plate 27 - 0.8%C steel microstructure - water quenched from 860°C, then tempered at 550°C for 30 minutes.

100μm

Plate 28 - SP 70 self-propelled 155 mm gun - with muzzle brake.

Plate 29 - AS 90 self-propelled 155 mm gun. *VSEL*

Plate 30 - SP 70 muzzle brake.

Plate 31 - M107 SP 175 mm gun barrel.

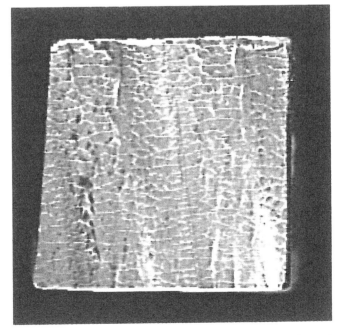

6mm

Plate 32 - Craze cracking on working surface of a 120 mm barrel section.

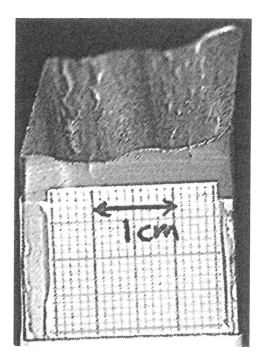

Plate 33 - Craze cracking section -
fatigue cracks growing
from the rifling roots.

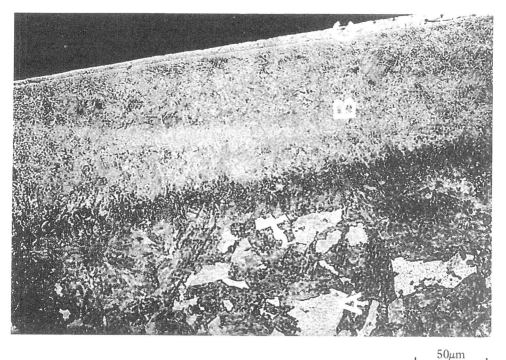

50μm

Plate 34 - Microstructure of working surface of fired gun barrel - transverse section,
optical micrograph.

Plate 35 - Microstructure of working surface of fired gun barrel - transverse section, SEM micrograph.

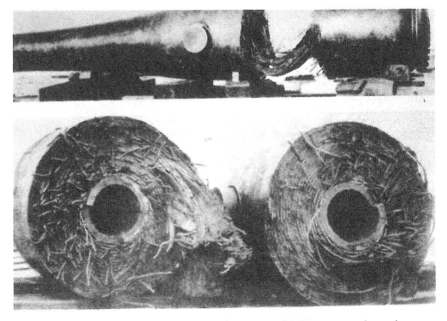

Plate 36 - Fracture of an old 'composite' wire wound 10" cannon barrel.

Plate 37 - 105 mm armour piercing
discarding sabot kinetic
energy penetrator round -
APDS KE round - sectioned.

Plate 38 - 120 mm armour piercing fin stabilised discarding sabot kinetic energy
penetrator round - APFSDS KE round.

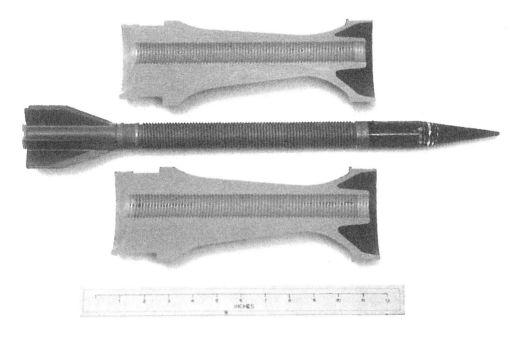

Plate 39 - 120 mm APFSDS KE penetrator round - sabots separated.

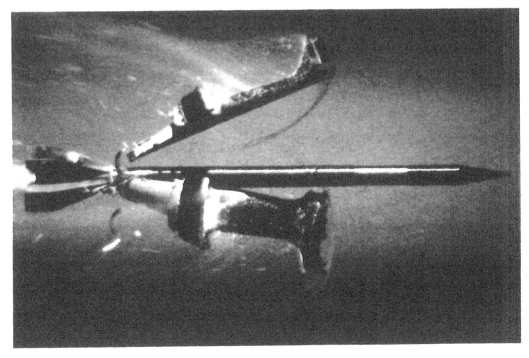

Plate 40 - Fired APFSDS soon after muzzle exit - sabots stripping away.

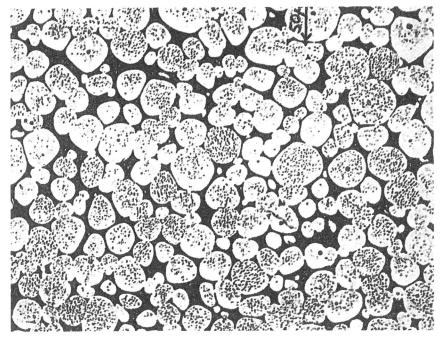

Plate 41 - Microstructure of W-10%Ni,Fe penetrator alloy.

$100\mu m$

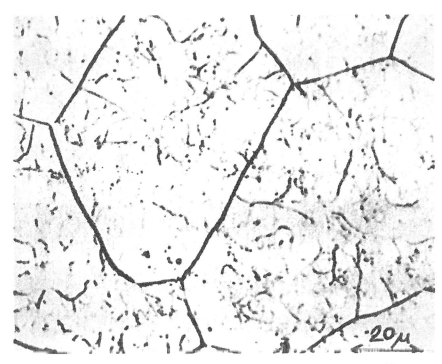

Plate 42 - Microstructure of DU penetrator alloy.

$20\mu m$

Plate 43 - Flash X-radiograph series - hydrodynamic penetration of a copper rod into an aluminium alloy target plate.

Plate 44 - LAW 80 shaped charge anti-tank weapon system. *Hunting Engineering*

Plate 45 - Mild steel target plates (each 25 mm thick) penetrated by a LAW 80 shaped charge jet. *Hunting Engineering*

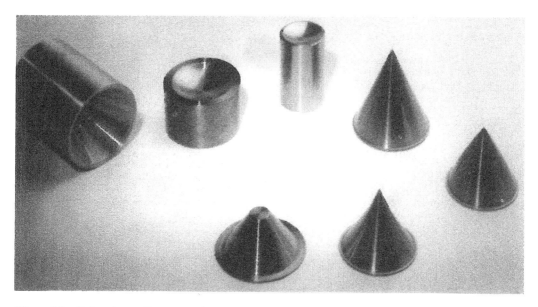

Plate 46 - Selection of copper shaped charge conical liners. *Hunting Engineering*

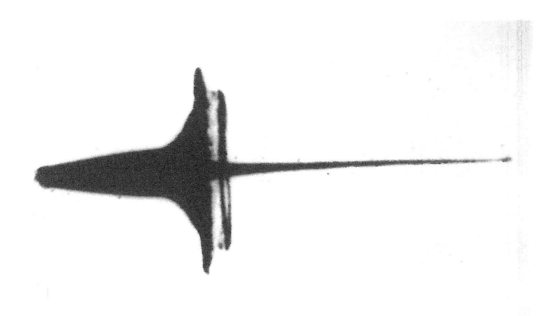

Plate 47 - Flash X-radiograph of copper cone hydrodynamic collapse into a jet.

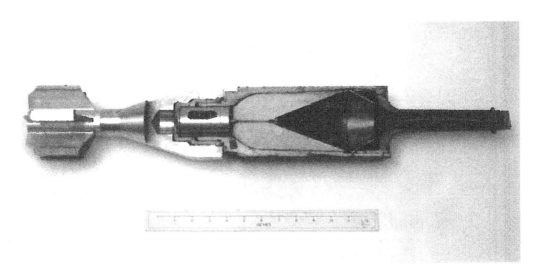

Plate 48 - Experimental 120 mm tank launched shaped charge warhead.

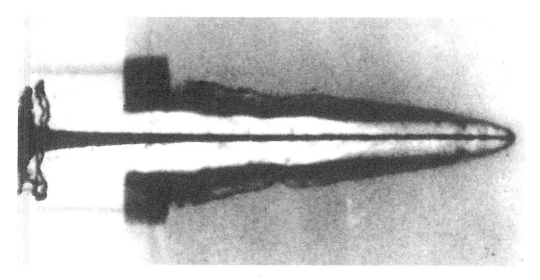

Plate 49 - Flash X-radiograph of copper jet penetrating hydrodynamically into an aluminium alloy target.

Plate 50 - 81 mm mortar.

Plate 51 - 81 mm mortar bomb body - cast iron.

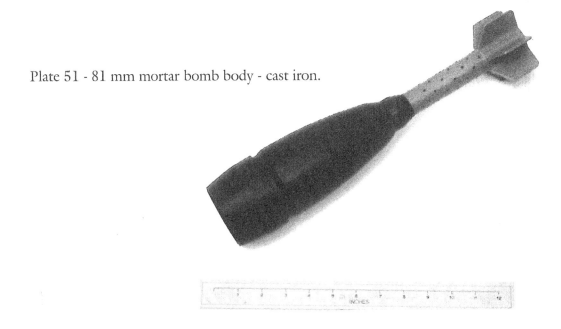

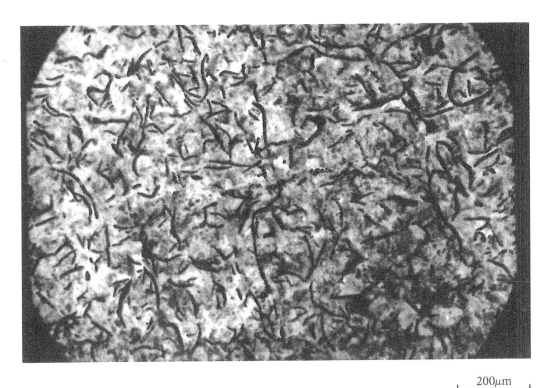

200µm

Plate 52 - Flake grey (automobile) cast iron microstructure.

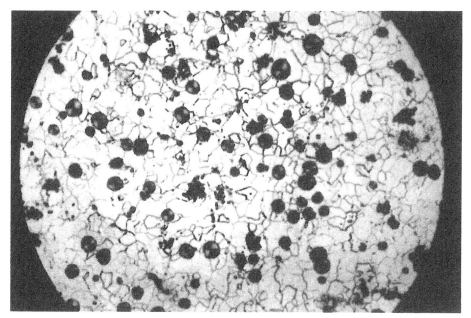

Plate 53 - Spheroidal graphite (sg) cast iron microstructure.  200μm

Plate 54 - 155mm high explosive
(HE) steel shell -
fragmenting type.

Plate 55 - Challenger main battle tank MBT - low alloy steel armour.

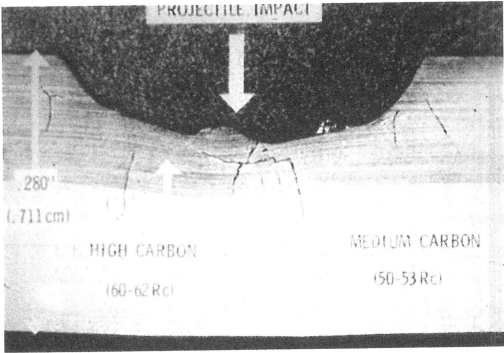

Plate 56 - Through-thickness section of face hardened steel armour plate after small calibre KE attack.

Plate 57 - Through-thickness section of steel plate penetrated by long rod
KE - curvature of tract due to obliquity.

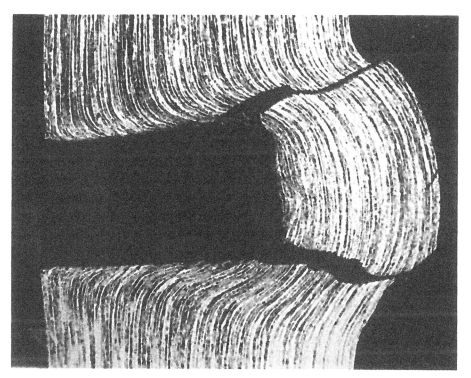

Plate 58 - Armour failure by 'plugging' - macrosection (aluminium alloy).

Plate 59 - 'Gross cracking' of a 50 mm thick low alloy steel plate.

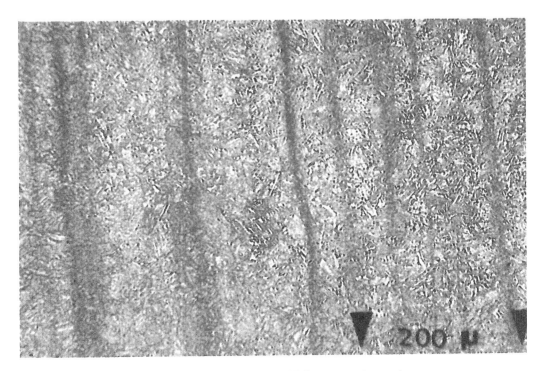

Plate 60 - 3%NiCrMo steel plate - through-thickness section microstructure.

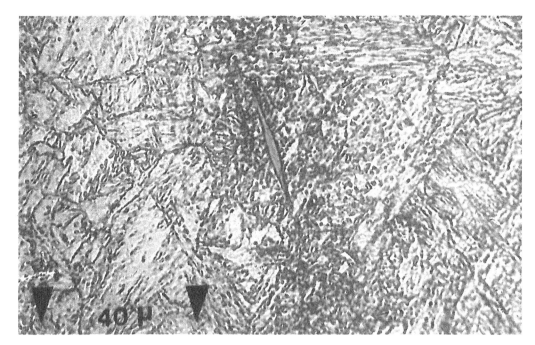

Plate 61 - 3%NiCrMo steel plate - through-thickness section microstructure at higher magnification.

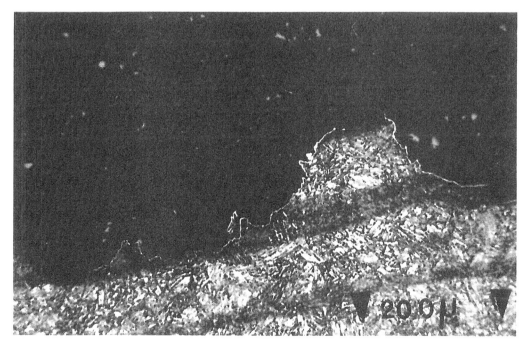

Plate 62 - 3%NiCrMo steel plate - section through fracture surface of through-thickness Charpy impact specimen, after testing at room temperature.

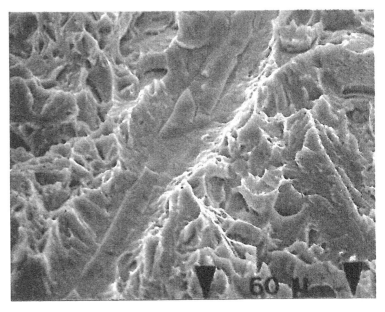

30μm

Plate 63 - 3%NiCrMo steel plate - SEM fractograph of through-thickness Charpy
impact specimen, after testing at minus 196°C.

Plate 64 - Electoslag remelted
ESR 3%NiCrMo
steel plate - through-
thickness section
microstructure.

electroslag remelting

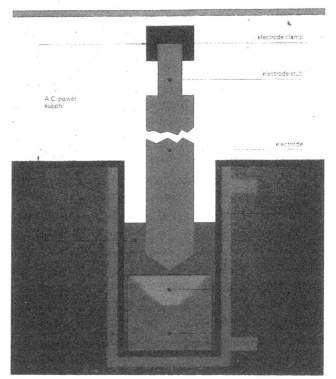

Plate 65 - Diagram of the ESR process. *Stocksbridge Engineering Steels*

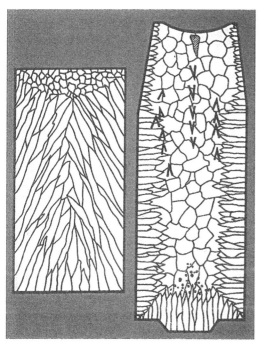

Plate 66 - Diagram of ingot cross-section macrostructures - ESR left, and air melted right.

1m

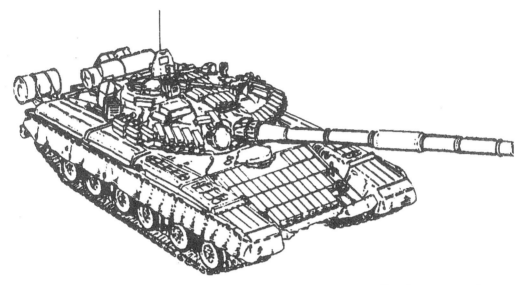

Plate 67 - Diagram of explosive reactive armour boxes (ERA) fitted onto a main battle tank (applique armour).

Plate 68 - M113 armoured personnel carrier APC - aluminium alloy armour.

Plate 69 - M113 armoured personnel carrier APC aluminium alloy armour plate - microstructural montage of the 3 principal planes.

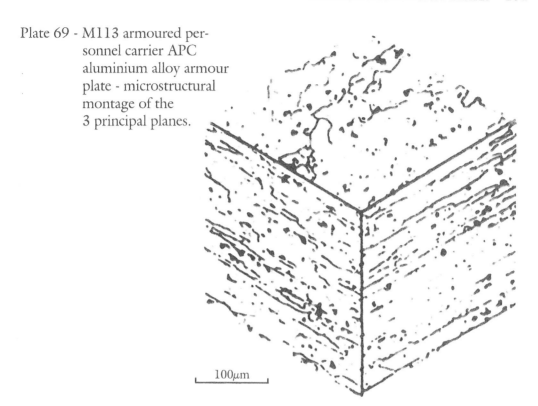

100μm

Plate 70 - Scorpion combat vehicle reconnaissance (tracked) CVR(T) - aluminium alloy armour.

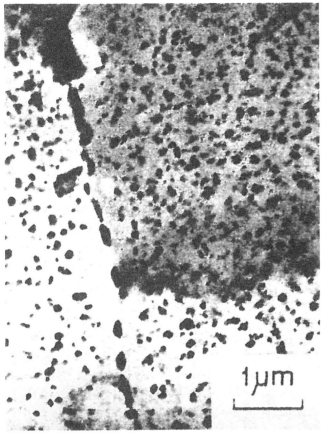

Plate 71 - Precipitation hardened aluminium alloy - SEM electron micrograph.

Plate 72 - Scorpion CVR(T) - showing 'buttering' of plate edges.

Plate 73 - Warrior infantry fighting vehicle IFV - aluminium alloy armour.

Plate 74 - Bradley IFV - aluminium alloy armour.

Plate75 - Bailey bridge (in New Zealand) - mild steel.

Plate 76 - Heavy girder bridge (in Jersey) - mild steel.

Plate 77 - Medium girder bridge MGB (with Chieftain tank) - aluminium alloy.

Plate 78 - MGB man portable section.

Plate 79 - MGB - double storey construction.

Plate 80 -MGB fitted with deflection limiting spars.

Plate 81 -
BR 90 - aluminium alloy.

Plate 82 - BR 90, with tank crossing.

Plate 83 - Armoured vehicle launched bridge AVLB being deployed - maraging steel.

Plate 84 - AVLB bridgelayer crossing its own bridge.

Plate 85 - 105 mm light gun.

Plate 86 - 105 mm light gun,
clearer view of trail
legs - alloy steel.

Plate 87 - 155 mm FH 70 gun.

Plate 88 - 155mm ultra-lightweight field howitzer UFH - titanium alloy trail legs. *VSEL*

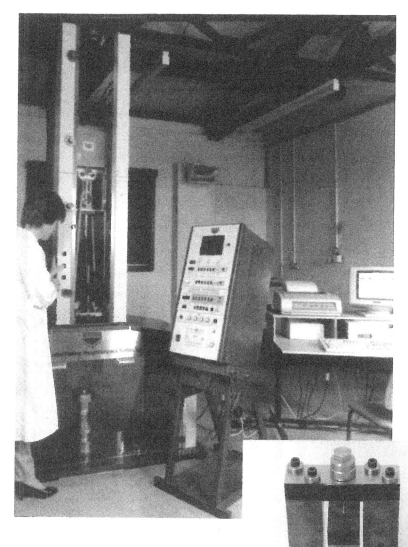

Plate 89 -
Instrumented drop
tower at RMCS.
Rosand

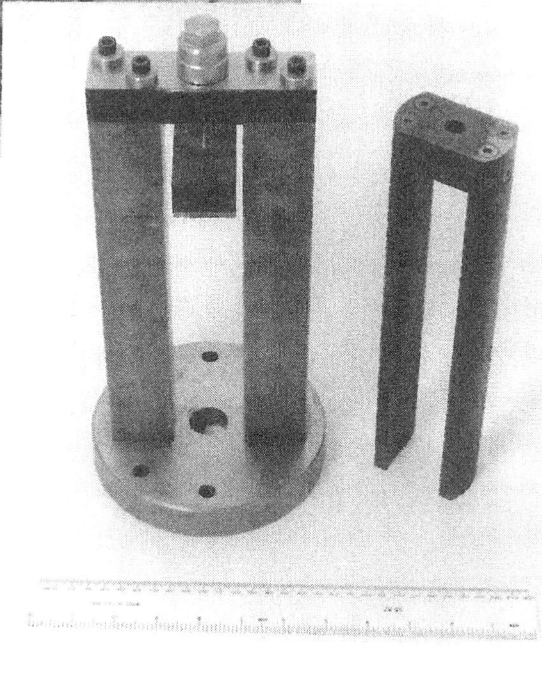

Plate 90 - Dynamic tensile rig
attachment

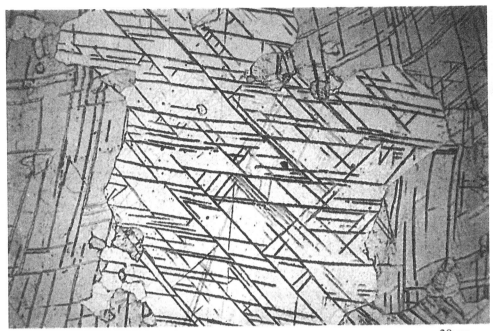

Plate 91 - Deformation twins in shock loaded iron (ferrite).   20μm

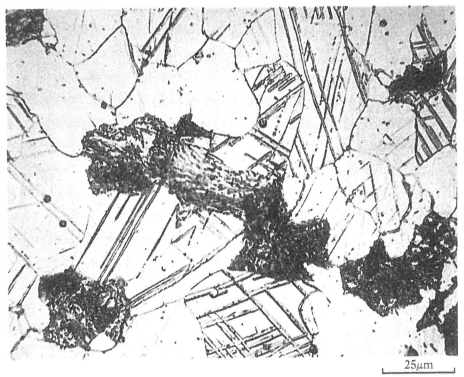

25μm

Plate 92 - Deformation twins in the ferrite grains of shock loaded mild steel.

Plate 93 - Adiabatic shear band in a medium carbon steel plate - after being partly penetrated by a kinetic energy KE round.

400μm

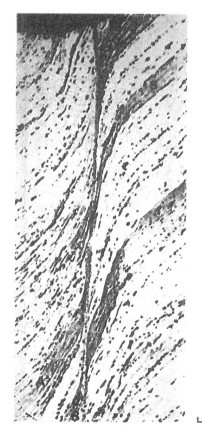

Plate 94 - Adiabatic shear band in a dynamically loaded aluminium alloy.

400μm

Plate 95 - Adiabatic shear band in a titanium alloy plate - after being partly penetrated by a KE round.

Plate 96 - Adiabatic shear band in a dynamically loaded DU alloy.

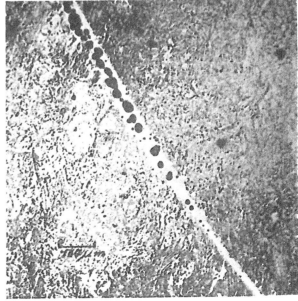

Index

Printed and bound by CPI Group (UK) Ltd, Croydon, CR0 4YY

23/10/2024

01777678-0014